Student Edition

Eureka Math
Grade 6
Modules 5 & 6

Special thanks go to the Gordon A. Cain Center and to the Department of Mathematics at Louisiana State University for their support in the development of *Eureka Math*.

Lesson 1: The Area of Parallelograms Through Rectangle Facts

Classwork

Opening Exercise

Name each shape.

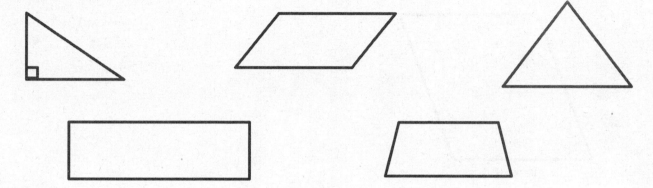

Exercises

1. Find the area of each parallelogram below. Note that the figures are not drawn to scale.

    a.

    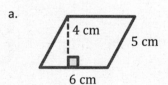

    4 cm
    5 cm
    6 cm

    b.

    8 m
    10 m
    25 m

    c.

    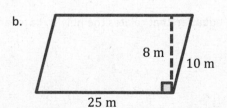

    7 ft.
    11.5 ft.
    12 ft.

2. Draw and label the height of each parallelogram. Use the correct mathematical tool to measure (in inches) the base and height, and calculate the area of each parallelogram.

    a.

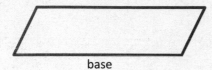

    base

    b.

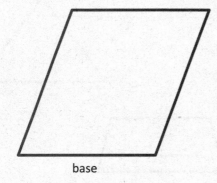

    base

    c.

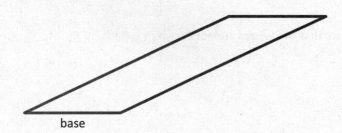

    base

3. If the area of a parallelogram is $\frac{35}{42}$ cm^2 and the height is $\frac{1}{7}$ cm, write an equation that relates the height, base, and area of the parallelogram. Solve the equation.

EUREKA
MATH™

Lesson Summary

The formula to calculate the area of a parallelogram is $A = bh$, where b represents the base and h represents the height of the parallelogram.

The height of a parallelogram is the line segment perpendicular to the base. The height is usually drawn from a vertex that is opposite the base.

Problem Set

Draw and label the height of each parallelogram.

1.

base

2.

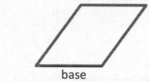

base

Calculate the area of each parallelogram. The figures are not drawn to scale.

3.

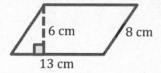

6 cm   8 cm

13 cm

4.

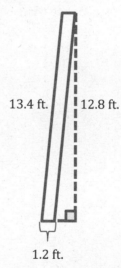

13.4 ft.   12.8 ft.

1.2 ft.

5.

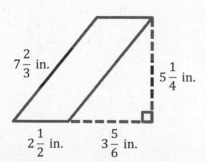

$7\frac{2}{3}$ in. $5\frac{1}{4}$ in.

$2\frac{1}{2}$ in. $3\frac{5}{6}$ in.

6.

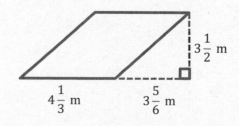

$3\frac{1}{2}$ m

$4\frac{1}{3}$ m $3\frac{5}{6}$ m

---

7. Brittany and Sid were both asked to draw the height of a parallelogram. Their answers are below.

Brittany   Sid

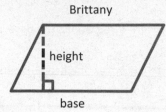   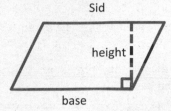

Are both Brittany and Sid correct? If not, who is correct? Explain your answer.

8. Do the rectangle and parallelogram below have the same area? Explain why or why not.

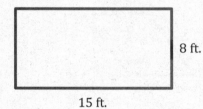   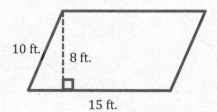

9. A parallelogram has an area of 20.3 cm^2 and a base of 2.5 cm. Write an equation that relates the area to the base and height, h. Solve the equation to determine the height of the parallelogram.

©2015 Great Minds eureka-math.org
G6-M5M6-SE-B3-1.3.1-01.2016

Lesson 2: The Area of Right Triangles

Classwork

Exploratory Challenge

a. Use the shapes labeled with an X to predict the formula needed to calculate the area of a right triangle. Explain your prediction.

Formula for the area of right triangles: _____

Area of the given triangle: _____

b. Use the shapes labeled with a Y to determine if the formula you discovered in part (a) is correct.

Does your area formula for triangle Y match the formula you got for triangle X?

If so, do you believe you have the correct formula needed to calculate the area of a right triangle? Why or why not?

If not, which formula do you think is correct? Why?

Area of the given triangle: _____

Exercises

Calculate the area of each triangle below. Each figure is not drawn to scale.

1.

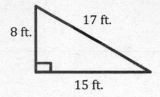

8 ft.   17 ft.

15 ft.

2.

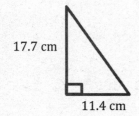

17.7 cm

11.4 cm

3.

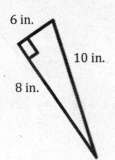

6 in.

10 in.

8 in.

4.

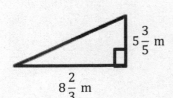

$5\frac{3}{5}$ m

$8\frac{2}{3}$ m

©2015 Great Minds eureka-math.org
G6-M5M6-SE-B3-1.3.1-01.2016

EUREKA MATH™

5.

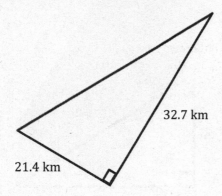

21.4 km

32.7 km

6. Mr. Jones told his students they each need half of a piece of paper. Calvin cut his piece of paper horizontally, and Matthew cut his piece of paper diagonally. Which student has the larger area on his half piece of paper? Explain.

Calvin's Paper

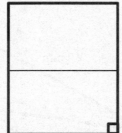

Matthew's Paper

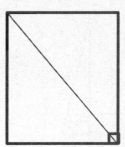

7. Ben requested that the rectangular stage be split into two equal sections for the upcoming school play. The only instruction he gave was that he needed the area of each section to be half of the original size. If Ben wants the stage to be split into two right triangles, did he provide enough information? Why or why not?

8. If the area of a right triangle is 6.22 sq. in. and its base is 3.11 in., write an equation that relates the area to the height, h, and the base. Solve the equation to determine the height.

©2015 Great Minds eureka-math.org
G6-M5M6-SE-B3-1.3.1-01.2016

Problem Set

Calculate the area of each right triangle below. Note that the figures are not drawn to scale.

1.
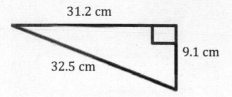

2.
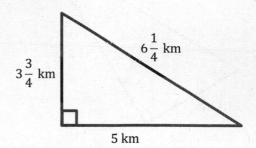

3.
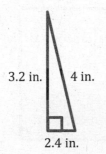

4.
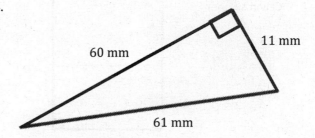

5.
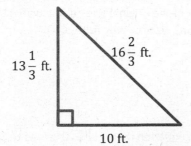

EUREKA
MATH™

6. Elania has two congruent rugs at her house. She cut one vertically down the middle, and she cut diagonally through the other one.

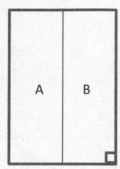   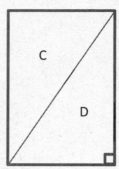

    After making the cuts, which rug (labeled A, B, C, or D) has the larger area? Explain.

7. Give the dimensions of a right triangle and a parallelogram with the same area. Explain how you know.

8. If the area of a right triangle is $\frac{9}{16}$ sq. ft. and the height is $\frac{3}{4}$ ft., write an equation that relates the area to the base, b, and the height. Solve the equation to determine the base.

This page intentionally left blank

Lesson 3: The Area of Acute Triangles Using Height and Base

Classwork

Exercises

1. Work with a partner on the exercises below. Determine if the area formula $A = \frac{1}{2}bh$ is always correct. You may use a calculator, but be sure to record your work on your paper as well. Figures are not drawn to scale.

	Area of Two Right Triangles	Area of Entire Triangle
15 cm, 17.4 cm, 12 cm, 9 cm, 12.6 cm triangle		
6.5 ft., 5.2 ft., 8 ft., 3.9 ft. triangle		
$2\frac{5}{6}$ in., 2 in., $\frac{5}{6}$ in. triangle		
34 m, 12 m, 32 m triangle		

2. Can we use the formula $A = \frac{1}{2} \times$ base $\times$ height to calculate the area of triangles that are not right triangles? Explain your thinking.

3. Examine the given triangle and expression.

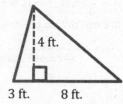

$$\frac{1}{2}(11 \text{ ft.})(4 \text{ ft.})$$

4 ft.

3 ft.   8 ft.

Explain what each part of the expression represents according to the triangle.

4. Joe found the area of a triangle by writing $A = \frac{1}{2}(11 \text{ in.})(4 \text{ in.})$, while Kaitlyn found the area by writing $A = \frac{1}{2}(3 \text{ in.})(4 \text{ in.}) + \frac{1}{2}(8 \text{ in.})(4 \text{ in.})$. Explain how each student approached the problem.

5. The triangle below has an area of 4.76 sq. in. If the base is 3.4 in., let h be the height in inches.

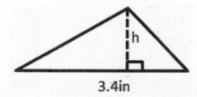

h

3.4in

a. Explain how the equation $4.76 \text{ in}^2 = \frac{1}{2}(3.4 \text{ in.})h$ represents the situation.

b. Solve the equation.

©2015 Great Minds eureka-math.org
G6-M5M6-SE-B3-1.3.1-01.2016

Problem Set

Calculate the area of each shape below. Figures are not drawn to scale.

1.

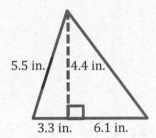

5.5 in.   4.4 in.

3.3 in.   6.1 in.

2.

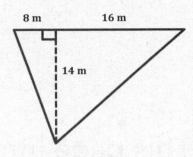

8 m   16 m

14 m

3.

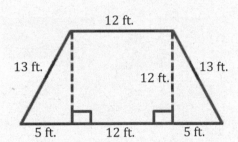

12 ft.

13 ft.   13 ft.

12 ft.

5 ft.   12 ft.   5 ft.

4.

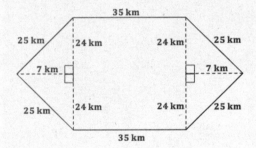

35 km

25 km   24 km   24 km   25 km

7 km   7 km

25 km   24 km   24 km   25 km

35 km

5. Immanuel is building a fence to make an enclosed play area for his dog. The enclosed area will be in the shape of a triangle with a base of 48 m. and an altitude of 32 m. How much space does the dog have to play?

6. Chauncey is building a storage bench for his son's playroom. The storage bench will fit into the corner and against two walls to form a triangle. Chauncey wants to buy a triangular shaped cover for the bench.

If the storage bench is $2\frac{1}{2}$ ft. along one wall and $4\frac{1}{4}$ ft. along the other wall, how big will the cover have to be to cover the entire bench?

Note: Figure is not to scale.

7. Examine the triangle to the right.

    a. Write an expression to show how you would calculate the area.

    b. Identify each part of your expression as it relates to the triangle.

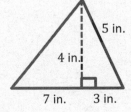

5 in.

4 in.

7 in.   3 in.

8. The floor of a triangular room has an area of $32\frac{1}{2}$ sq. m. If the triangle's altitude is $7\frac{1}{2}$ m, write an equation to determine the length of the base, b, in meters. Then solve the equation.

This page intentionally left blank

Lesson 4: The Area of All Triangles Using Height and Base

Classwork

Opening Exercise

Draw and label the altitude of each triangle below.

a.

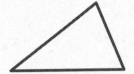

b.

c.

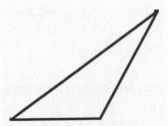

Exploratory Challenge/Exercises 1–5

1. Use rectangle X and the triangle with the altitude inside (triangle X) to show that the area formula for the triangle is $A = \frac{1}{2} \times$ base $\times$ height.

    a. Step One: Find the area of rectangle X.

    b. Step Two: What is half the area of rectangle X?

c. Step Three: Prove, by decomposing triangle X, that it is the same as half of rectangle X. Please glue your decomposed triangle onto a separate sheet of paper. Glue it into rectangle X. What conclusions can you make about the triangle's area compared to the rectangle's area?

2. Use rectangle Y and the triangle with a side that is the altitude (triangle Y) to show the area formula for the triangle is $A = \frac{1}{2} \times$ base $\times$ height.

a. Step One: Find the area of rectangle Y.

b. Step Two: What is half the area of rectangle Y?

c. Step Three: Prove, by decomposing triangle Y, that it is the same as half of rectangle Y. Please glue your decomposed triangle onto a separate sheet of paper. Glue it into rectangle Y. What conclusions can you make about the triangle's area compared to the rectangle's area?

3. Use rectangle Z and the triangle with the altitude outside (triangle Z) to show the area formula for the triangle is $A = \frac{1}{2} \times$ base $\times$ height.

a. Step One: Find the area of rectangle Z.

b. Step Two: What is half the area of rectangle Z?

c. Step Three: Prove, by decomposing triangle Z, that it is the same as half of rectangle Z. Please glue your decomposed triangle onto a separate sheet of paper. Glue it into rectangle Z. What conclusions can you make about the triangle's area compared to the rectangle's area?

4. When finding the area of a triangle, does it matter where the altitude is located?

5. How can you determine which part of the triangle is the base and which is the height?

Exercises 6–8

Calculate the area of each triangle. Figures are not drawn to scale.

6.

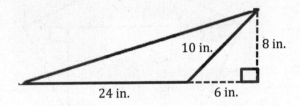

7.

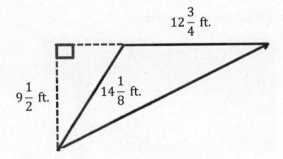

8. Draw three triangles (acute, right, and obtuse) that have the same area. Explain how you know they have the same area.

Problem Set

Calculate the area of each figure below. Figures are not drawn to scale.

1.

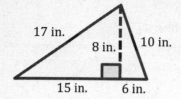
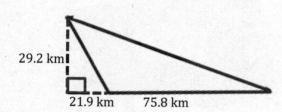

17 in.   8 in.   10 in.

15 in.   6 in.

2.

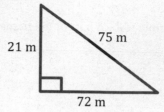

75 m

21 m

72 m

3.

29.2 km

21.9 km   75.8 km

4.

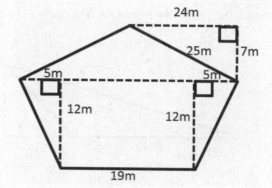

24m

25m   7m

5m   5m

12m   12m

19m

5. The Andersons are going on a long sailing trip during the summer. However, one of the sails on their sailboat ripped, and they have to replace it. The sail is pictured below.

If the sailboat sails are on sale for $2 per square foot, how much will the new sail cost?

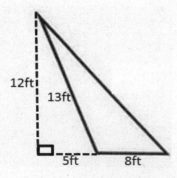

12ft

13ft

5ft   8ft

EUREKA
MATH™

6.   Darnell and Donovan are both trying to calculate the area of an obtuse triangle. Examine their calculations below.

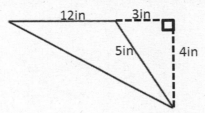

Darnell's Work	Donovan's Work
$A = \dfrac{1}{2} \times 3 \text{ in.} \times 4 \text{ in.}$ $A = 6 \text{ in}^2$	$A = \dfrac{1}{2} \times 12 \text{ in.} \times 4 \text{ in.}$ $A = 24 \text{ in}^2$

Which student calculated the area correctly? Explain why the other student is not correct.

7.   Russell calculated the area of the triangle below. His work is shown.

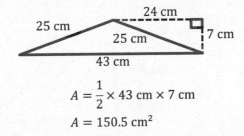

$$A = \frac{1}{2} \times 43 \text{ cm} \times 7 \text{ cm}$$
$$A = 150.5 \text{ cm}^2$$

Although Russell was told his work is correct, he had a hard time explaining why it is correct. Help Russell explain why his calculations are correct.

8.   The larger triangle below has a base of 10.14 m; the gray triangle has an area of 40.325 m².

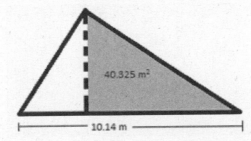

a.   Determine the area of the larger triangle if it has a height of 12.2 m.

b.   Let A be the area of the unshaded (white) triangle in square meters. Write and solve an equation to determine the value of A, using the areas of the larger triangle and the gray triangle.

This page intentionally left blank

Lesson 5: The Area of Polygons Through Composition and Decomposition

Classwork

Opening Exercise

Here is an aerial view of a woodlot.

If $AB = 10$ units, $FE = 8$ units, $AF = 6$ units, and $DE = 7$ units, find the lengths of the other two sides.

$DC =$

$BC =$

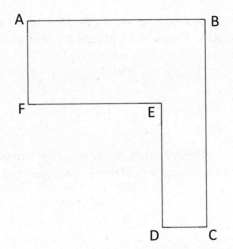

If $DC = 10$ units, $FE = 30$ units, $AF = 28$ units, and $BC = 54$ units, find the lengths of the other two sides.

$AB =$

$DE =$

Discussion

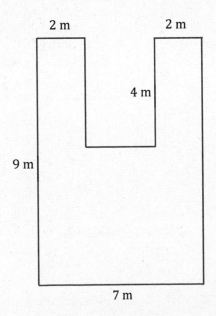

Example 1: Decomposing Polygons into Rectangles

The Intermediate School is producing a play that needs a special stage built. A diagram of the stage is shown below (not to scale).

a. On the first diagram, divide the stage into three rectangles using two horizontal lines. Find the dimensions of these rectangles, and calculate the area of each. Then, find the total area of the stage.

b. On the second diagram, divide the stage into three rectangles using two vertical lines. Find the dimensions of these rectangles, and calculate the area of each. Then, find the total area of the stage.

c. On the third diagram, divide the stage into three rectangles using one horizontal line and one vertical line. Find the dimensions of these rectangles, and calculate the area of each. Then, find the total area of the stage.

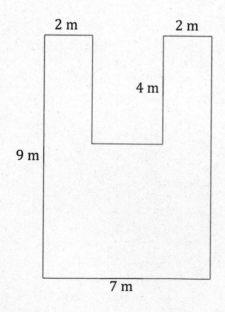
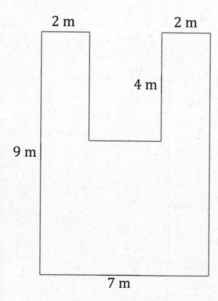
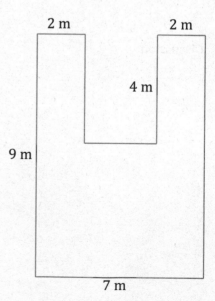

EUREKA
MATH™

d. Think of this as a large rectangle with a piece removed.

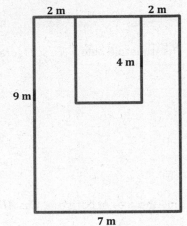

i. What are the dimensions of the large rectangle and the small rectangle?

ii. What are the areas of the two rectangles?

iii. What operation is needed to find the area of the original figure?

iv. What is the difference in area between the two rectangles?

v. What do you notice about your answers to (a), (b), (c), and (d)?

vi. Why do you think this is true?

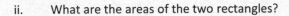

Example 2: Decomposing Polygons into Rectangles and Triangles

Parallelogram $ABCD$ is part of a large solar power collector. The base measures 6 m and the height is 4 m.

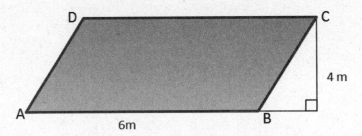

a. Draw a diagonal from A to C. Find the area of both triangles ABC and ACD.

b. Draw in the other diagonal, from B to D. Find the area of both triangles ABD and BCD.

Example 3: Decomposing Trapezoids

The trapezoid below is a scale drawing of a garden plot.

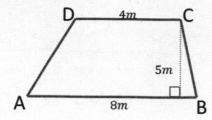

Find the area of both triangles ABC and ACD. Then find the area of the trapezoid.

Find the area of both triangles ABD and BCD. Then find the area of the trapezoid.

How else could we find this area?

EUREKA
MATH

Problem Set

1. If $AB = 20$ units, $FE = 12$ units, $AF = 9$ units, and $DE = 12$ units, find the length of the other two sides. Then, find the area of the irregular polygon.

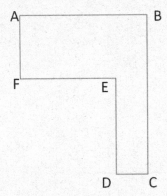

2. If $DC = 1.9$ cm, $FE = 5.6$ cm, $AF = 4.8$ cm, and $BC = 10.9$ cm, find the length of the other two sides. Then, find the area of the irregular polygon.

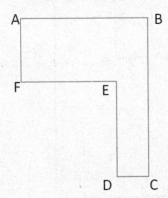

3. Determine the area of the trapezoid below. The trapezoid is not drawn to scale.

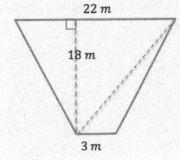

EUREKA MATH™

©2015 Great Minds eureka-math.org
G6-M5M6-SE-B3-1.3.1-01.2016

4. Determine the area of the shaded isosceles trapezoid below. The image is not drawn to scale.

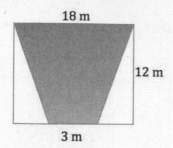

5. Here is a sketch of a wall that needs to be painted:

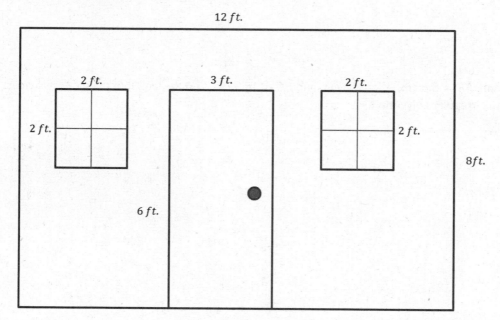

a. The windows and door will not be painted. Calculate the area of the wall that will be painted.

b. If a quart of Extra-Thick Gooey Sparkle paint covers 30 ft^2, how many quarts must be purchased for the painting job?

6. The figure below shows a floor plan of a new apartment. New carpeting has been ordered, which will cover the living room and bedroom but not the kitchen or bathroom. Determine the carpeted area by composing or decomposing in two different ways, and then explain why they are equivalent.

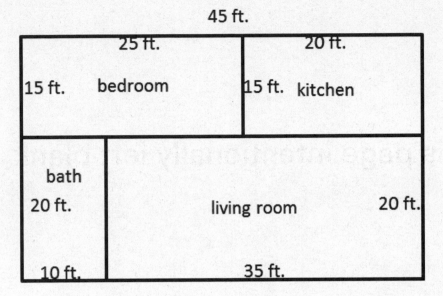

This page intentionally left blank

Lesson 6: Area in the Real World

Classwork

Exploratory Challenge 1: Classroom Wall Paint

The custodians are considering painting our classroom next summer. In order to know how much paint they must buy, the custodians need to know the total surface area of the walls. Why do you think they need to know this, and how can we find the information?

Make a prediction of how many square feet of painted surface there are on one wall in the room. If the floor has square tiles, these can be used as a guide.

Estimate the dimensions and the area. Predict the area before you measure.

My prediction: _____ ft^2.

    a.   Measure and sketch one classroom wall. Include measurements of windows, doors, or anything else that would not be painted.

        Sketch:

Object or Item to Be Measured	Measurement Units	Precision (measure to the nearest)	Length	Width	Expression that Shows the Area	Area
door	feet	half foot	$6\frac{1}{2}$ ft.	$3\frac{1}{2}$ ft.	$6\frac{1}{2}$ ft. $\times$ $3\frac{1}{2}$ ft.	$22\frac{3}{4}$ ft^2

b. Work with your partners and your sketch of the wall to determine the area that needs paint. Show your sketch and calculations below; clearly mark your measurements and area calculations.

c. A gallon of paint covers about 350 ft^2. Write an expression that shows the total area of the wall. Evaluate it to find how much paint is needed to paint the wall.

d. How many gallons of paint would need to be purchased to paint the wall?

EUREKA MATH™

Exploratory Challenge 2

Object or Item to Be Measured	Measurement Units	Precision (measure to the nearest)	Length	Width	Area
door	feet	half foot	$6\frac{1}{2}$ ft.	$3\frac{1}{2}$ ft.	$22\frac{3}{4}$ ft²

Problem Set

1. Below is a drawing of a wall that is to be covered with either wallpaper or paint. The wall is 8 ft. high and 16 ft. wide. The window, mirror, and fireplace are not to be painted or papered. The window measures 18 in. wide and 14 ft. high. The fireplace is 5 ft. wide and 3 ft. high, while the mirror above the fireplace is 4 ft. wide and 2 ft. high. (Note: this drawing is not to scale.)

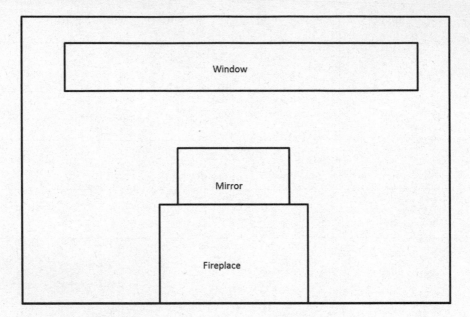

a. How many square feet of wallpaper are needed to cover the wall?

b. The wallpaper is sold in rolls that are 18 in. wide and 33 ft. long. Rolls of solid color wallpaper will be used, so patterns do not have to match up.

   i.   What is the area of one roll of wallpaper?

   ii. How many rolls would be needed to cover the wall?

c. This week, the rolls of wallpaper are on sale for $11.99/roll. Find the cost of covering the wall with wallpaper.

d. A gallon of special textured paint covers 200 ft^2 and is on sale for $22.99/gallon. The wall needs to be painted twice (the wall needs two coats of paint). Find the cost of using paint to cover the wall.

2. A classroom has a length of 30 ft. and a width of 20 ft. The flooring is to be replaced by tiles. If each tile has a length of 36 in. and a width of 24 in., how many tiles are needed to cover the classroom floor?

3. Challenge: Assume that the tiles from Problem 2 are unavailable. Another design is available, but the tiles are square, 18 in. on a side. If these are to be installed, how many must be ordered?

©2015 Great Minds eureka-math.org
G6-M5M6-SE-B3-1.3.1-01.2016

4. A rectangular flower bed measures 10 m by 6 m. It has a path 2 m wide around it. Find the area of the path.

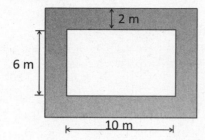

5. A diagram of Tracy's deck is shown below, shaded blue. He wants to cover the missing portion of his deck with soil in order to grow a garden.

    a. Find the area of the missing portion of the deck. Write the expression and evaluate it.

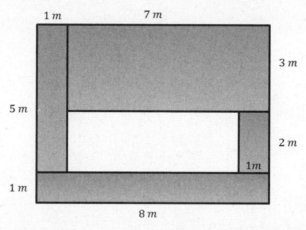

    b. Find the missing portion of the deck using a different method. Write the expression and evaluate it.

    c. Write two equivalent expressions that can be used to determine the area of the missing portion of the deck.

    d. Explain how each expression demonstrates a different understanding of the diagram.

6. The entire large rectangle below has an area of $3\frac{1}{2}$ ft². If the dimensions of the white rectangle are as shown below, write and solve an equation to find the area, A, of the shaded region.

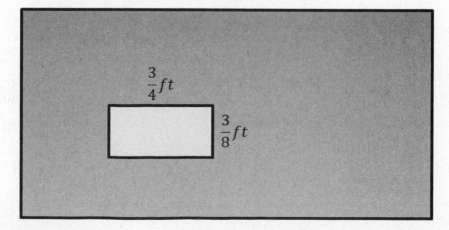

This page intentionally left blank

Lesson 7: Distance on the Coordinate Plane

Classwork

Example

Determine the lengths of the given line segments by determining the distance between the two endpoints.

Line Segment	Point	Point	Distance	Proof
$\overline{AB}$				
$\overline{BC}$				
$\overline{CD}$				
$\overline{BD}$				
$\overline{DE}$				
$\overline{EF}$				
$\overline{FG}$				
$\overline{EG}$				
$\overline{GA}$				
$\overline{FA}$				
$\overline{EA}$				

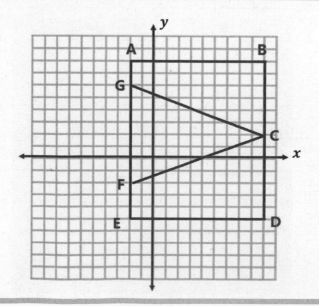

EUREKA MATH™

Exercise

Complete the table using the diagram on the coordinate plane.

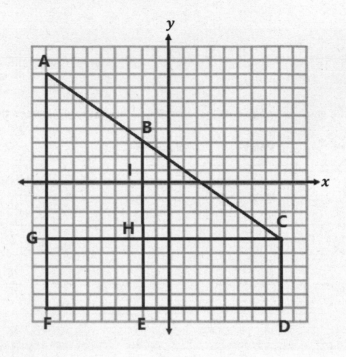

Line Segment	Point	Point	Distance	Proof
$\overline{BI}$				
$\overline{BH}$				
$\overline{BE}$				
$\overline{GH}$				
$\overline{HC}$				
$\overline{GC}$				
$\overline{CD}$				
$\overline{FG}$				
$\overline{GA}$				
$\overline{AF}$				

EUREKA
MATH™

Extension

For each problem below, write the coordinates of two points that are 5 units apart with the segment connecting these points having the following characteristics.

    a.   The segment is vertical.

    b.   The segment intersects the x-axis.

    c.   The segment intersects the y-axis.

    d.   The segment is vertical and lies above the x-axis.

©2015 Great Minds eureka-math.org
G6-M5M6-SE-B3-1.3.1-01.2016

Problem Set

1. Given the pairs of points, determine whether the segment that joins them is horizontal, vertical, or neither.

   a. $X(3,5)$ and $Y(-2,5)$ _____

   b. $M(-4,9)$ and $N(4,-9)$ _____

   c. $E(-7,1)$ and $F(-7,4)$ _____

2. Complete the table using absolute value to determine the lengths of the line segments.

Line Segment	Point	Point	Distance	Proof
$\overline{AB}$	$(-3,5)$	$(7,5)$		
$\overline{CD}$	$(1,-3)$	$(-6,-3)$		
$\overline{EF}$	$(2,-9)$	$(2,-3)$		
$\overline{GH}$	$(6,1)$	$(6,16)$		
$\overline{JK}$	$(-3,0)$	$(-3,12)$		

3. Complete the table using the diagram and absolute value to determine the lengths of the line segments.

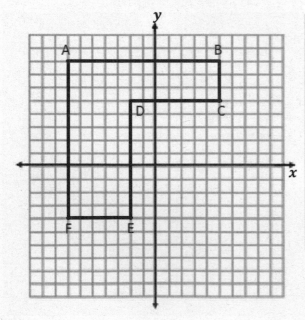

Line Segment	Point	Point	Distance	Proof
$\overline{AB}$				
$\overline{BC}$				
$\overline{CD}$				
$\overline{DE}$				
$\overline{EF}$				
$\overline{FA}$				

EUREKA
MATH™

©2015 Great Minds eureka-math.org
G6-M5M6-SE-B3-1.3.1-01.2016

4.   Complete the table using the diagram and absolute value to determine the lengths of the line segments.

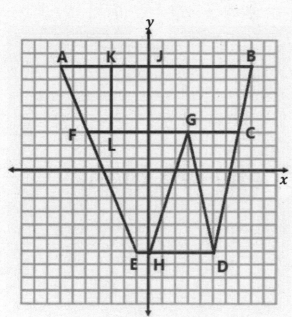

Line Segment	Point	Point	Distance	Proof
$\overline{AB}$				
$\overline{CG}$				
$\overline{CF}$				
$\overline{GF}$				
$\overline{DH}$				
$\overline{DE}$				
$\overline{HJ}$				
$\overline{KL}$				

5.   Name two points in different quadrants that form a vertical line segment that is 8 units in length.

6.   Name two points in the same quadrant that form a horizontal line segment that is 5 units in length.

©2015 Great Minds eureka-math.org
G6-M5M6-SE-B3-1.3.1-01.2016

This page intentionally left blank

Lesson 8: Drawing Polygons in the Coordinate Plane

Classwork

Examples

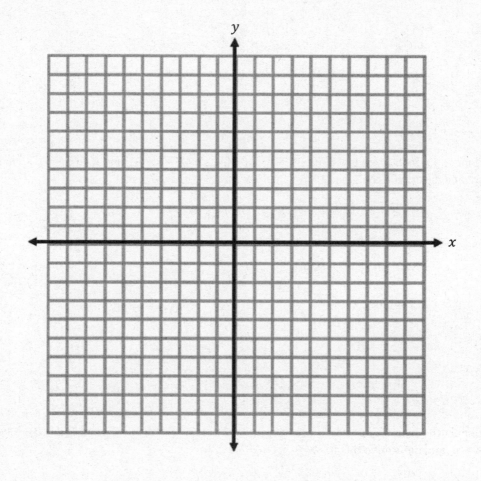

1. Plot and connect the points $A(3, 2)$, $B(3, 7)$, and $C(8, 2)$. Name the shape, and determine the area of the polygon.

EUREKA
MATH™

©2015 Great Minds eureka-math.org
G6-M5M6-SE-B3-1.3.1-01.2016

2. Plot and connect the points $E(-8, 8)$, $F(-2, 5)$, and $G(7, 2)$. Then give the best name for the polygon, and determine the area.

3. Plot and connect the following points: $K(-9, -7)$, $L(-4, -2)$, $M(-1, -5)$, and $N(-5, -5)$. Give the best name for the polygon, and determine the area.

4. Plot and connect the following points: $P(1, -4)$, $Q(5, -2)$, $R(9, -4)$, $S(7, -8)$, and $T(3, -8)$. Give the best name for the polygon, and determine the area.

EUREKA
MATH

©2015 Great Minds eureka-math.org
G6-M5M6-SE-B3-1.3.1-01.2016

5. Two of the coordinates of a rectangle are $A(3,7)$ and $B(3,2)$. The rectangle has an area of 30 square units. Give the possible locations of the other two vertices by identifying their coordinates. (Use the coordinate plane to draw and check your answer.)

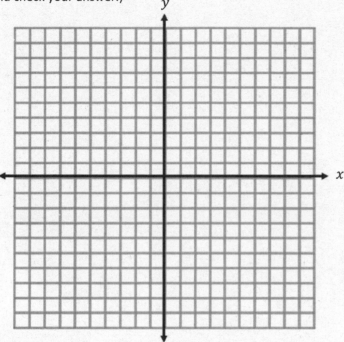

Exercises

For Exercises 1 and 2, plot the points, name the shape, and determine the area of the shape. Then write an expression that could be used to determine the area of the figure. Explain how each part of the expression corresponds to the situation.

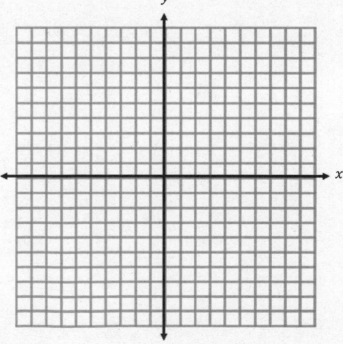

1. $A(4, 6)$, $B(8, 6)$, $C(10, 2)$, $D(8, -3)$, $E(5, -3)$, and $F(2, 2)$

2. $X(-9, 6)$, $Y(-2, -1)$, and $Z(-8, -7)$

3. A rectangle with vertices located at $(-3, 4)$ and $(5, 4)$ has an area of 32 square units. Determine the location of the other two vertices.

4. Challenge: A triangle with vertices located at $(-2, -3)$ and $(3, -3)$ has an area of 20 square units. Determine one possible location of the other vertex.

©2015 Great Minds eureka-math.org
G6-M5M6-SE-B3-1.3.1-01.2016

Problem Set

Plot the points for each shape, determine the area of the polygon, and then write an expression that could be used to determine the area of the figure. Explain how each part of the expression corresponds to the situation.

1. $A\,(1,3)$, $B\,(2,8)$, $C\,(8,8)$, $D\,(10,3)$, and $E\,(5,-2)$

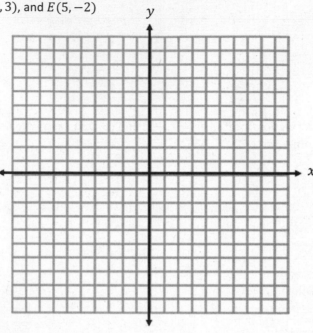

2. $X(-10,2)$, $Y(-3,6)$, and $Z(-6,-5)$

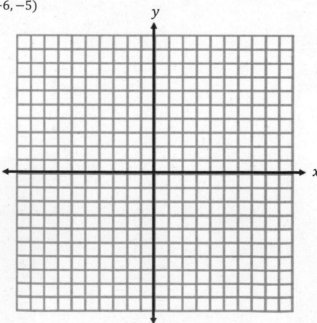

EUREKA
MATH™

©2015 Great Minds eureka-math.org
G6-M5M6-SE-B3-1.3.1-01.2016

3. $E(5, 7)$, $F(9, -5)$, and $G(1, -3)$

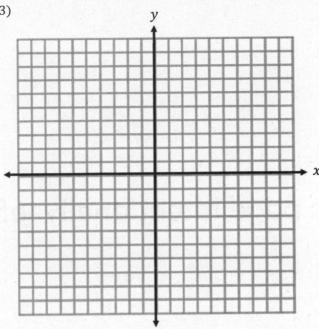

4. Find the area of the triangle in Problem 3 using a different method. Then, compare the expressions that can be used for both solutions in Problems 3 and 4.

5. Two vertices of a rectangle are $(8, -5)$ and $(8, 7)$. If the area of the rectangle is 72 square units, name the possible location of the other two vertices.

6. A triangle with two vertices located at $(5, -8)$ and $(5, 4)$ has an area of 48 square units. Determine one possible location of the other vertex.

This page intentionally left blank

Lesson 9: Determining Perimeter and Area of Polygons on the Coordinate Plane

Classwork

Example 1

Jasjeet has made a scale drawing of a vegetable garden she plans to make in her backyard. She needs to determine the perimeter and area to know how much fencing and dirt to purchase. Determine both the perimeter and area.

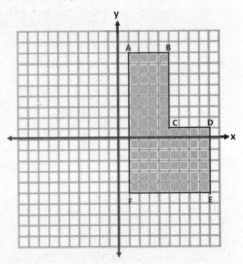

Example 2

Calculate the area of the polygon using two different methods. Write two expressions to represent the two methods, and compare the structure of the expressions.

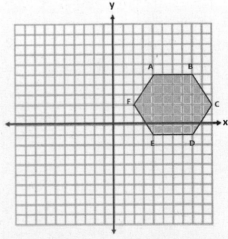

Exercises

1. Determine the area of the following shapes.

   a.

   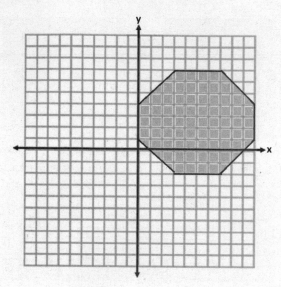

   b.

   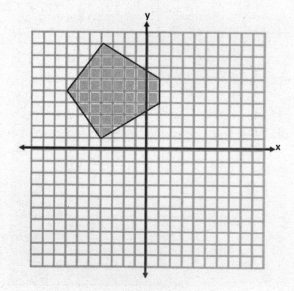

2. Determine the area and perimeter of the following shapes.

a.

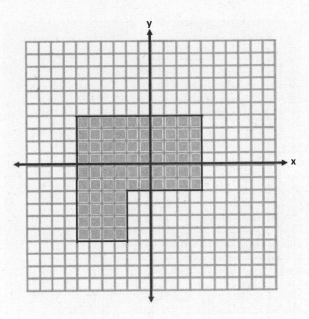

b.

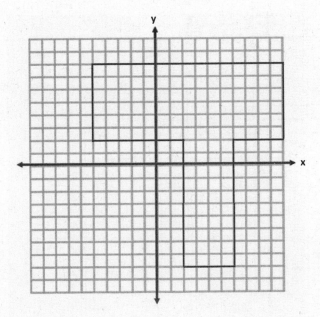

Problem Set

1. Determine the area of the polygon.

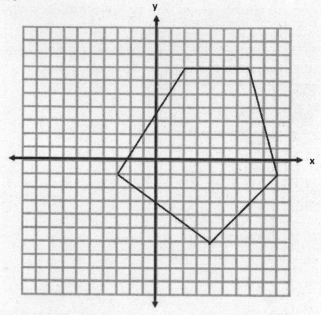

2. Determine the area and perimeter of the polygon.

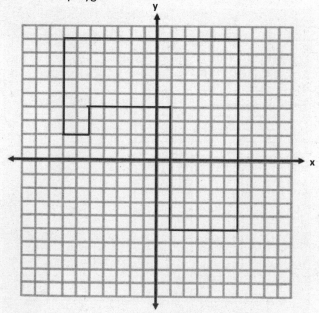

EUREKA
MATH™

3. Determine the area of the polygon. Then, write an expression that could be used to determine the area.

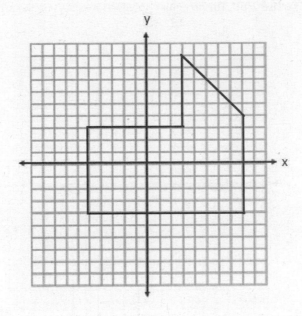

4. If the length of each square was worth 2 instead of 1, how would the area in Problem 3 change? How would your expression change to represent this area?

5. Determine the area of the polygon. Then, write an expression that represents the area.

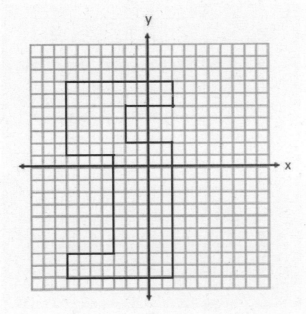

6. Describe another method you could use to find the area of the polygon in Problem 5. Then, state how the expression for the area would be different than the expression you wrote.

7. Write one of the letters from your name using rectangles on the coordinate plane. Then, determine the area and perimeter. (For help see Exercise 2(b). This irregular polygon looks sort of like a T.)

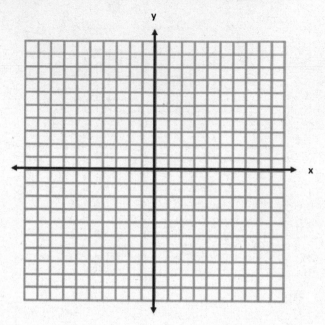

EUREKA
MATH™

Lesson 10: Distance, Perimeter, and Area in the Real World

Classwork

Opening Exercise

a. Find the area and perimeter of this rectangle:

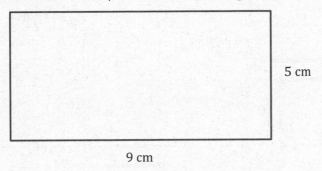

5 cm

9 cm

b. Find the width of this rectangle. The area is 1.2 m^2, and the length is 1.5 m.

$$A = 1.2\ m^2 \quad w = ?$$

$$l = 1.5\ m$$

Example: Student Desks or Tables

1. Measure the dimensions of the top of your desk.

2. How do you find the area of the top of your desk?

3. How do you find the perimeter?

4. Record these on your paper in the appropriate column.

©2015 Great Minds eureka-math.org
G6-M5M6-SE-B3-1.3.1-01.2016

Exploratory Challenge

Estimate and predict the area and perimeter of each object. Then measure each object, and calculate both the area and perimeter of each.

Object or Item to be Measured	Measurement Units	Precision (measure to the nearest)	Area Prediction (square units)	Area (square units) Write the expression and evaluate it.	Perimeter Prediction (linear units)	Perimeter (linear units)
Ex: door	feet	half foot		$6\frac{1}{2}$ ft. $\times 3\frac{1}{2}$ ft. $= 22\frac{3}{4}$ ft^2		$2\left(3\frac{1}{2}\text{ ft.} + 6\frac{1}{2}\text{ ft.}\right)$ $= 20$ ft.
desktop						

Optional Challenge

Object or Item to be Measured	Measurement Units	Precision (measure to the nearest)	Area (square units)	Perimeter (linear units)
Ex: door	feet	half foot	$6\frac{1}{2}$ ft. $\times$ $3\frac{1}{2}$ ft. $= 22\frac{3}{4}$ ft^2	$2\left(3\frac{1}{2}\text{ ft.} + 6\frac{1}{2}\text{ ft.}\right)$ $= 20$ ft.

©2015 Great Minds eureka-math.org
G6-M5M6-SE-B3-1.3.1-01.2016

Problem Set

1. How is the length of the side of a square related to its area and perimeter? The diagram below shows the first four squares stacked on top of each other with their upper left-hand corners lined up. The length of one side of the smallest square is 1 foot.

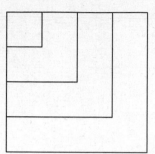

a. Complete this chart calculating area and perimeter for each square.

Side Length (in feet)	Expression Showing the Area	Area (in square feet)	Expression Showing the Perimeter	Perimeter (in feet)
1	1×1	1	1×4	4
2				
3				
4				
5				
6				
7				
8				
9				
10				
n				

b. In a square, which numerical value is greater, the area or the perimeter?

c. When is the numerical value of a square's area (in square units) equal to its perimeter (in units)?

d. Why is this true?

EUREKA MATH

©2015 Great Minds eureka-math.org
G6-M5M6-SE-B3-1.3.1-01.2016

2. This drawing shows a school pool. The walkway around the pool needs special nonskid strips installed but only at the edge of the pool and the outer edges of the walkway.

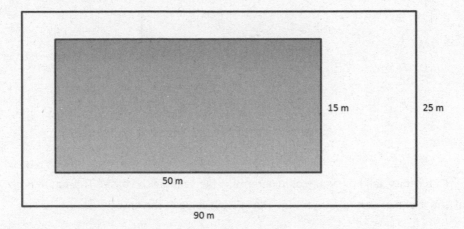

15 m   25 m

50 m

90 m

   a. Find the length of nonskid strips that is needed for the job.

   b. The nonskid strips are sold only in rolls of 50 m. How many rolls need to be purchased for the job?

3. A homeowner called in a painter to paint the walls and ceiling of one bedroom. His bedroom is 18 ft. long, 12 ft. wide, and 8 ft. high. The room has <u>two</u> doors, each 3 ft. by 7 ft., and <u>three</u> windows each 3 ft. by 5 ft. The doors and windows will not be painted. A gallon of paint can cover 300 ft^2. A hired painter claims he needs a minimum of 4 gallons. Show that his estimate is too high.

4. Theresa won a gardening contest and was awarded a roll of deer-proof fencing. The fencing is 36 feet long. She and her husband, John, discuss how to best use the fencing to make a rectangular garden. They agree that they should only use whole numbers of feet for the length and width of the garden.

   a. What are all of the possible dimensions of the garden?

   b. Which plan yields the maximum area for the garden? Which plan yields the minimum area?

5. Write and then solve the equation to find the missing value below.

$$A = 1.82\ m^2 \qquad w = ?$$

$$l = 1.4\ m$$

6.   Challenge: This is a drawing of the flag of the Republic of the Congo. The area of this flag is $3\frac{3}{4}$ ft².

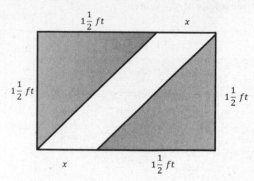

a.   Using the area formula, tell how you would determine the value of the base. This figure is not drawn to scale.

b.   Using what you found in part (a), determine the missing value of the base.

Lesson 11: Volume with Fractional Edge Lengths and Unit Cubes

Classwork

Opening Exercise

Which prism holds more 1 in. × 1 in. × 1 in. cubes? How many more cubes does the prism hold?

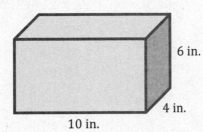

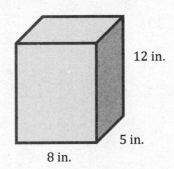

Example 1

A box with the same dimensions as the prism in the Opening Exercise is used to ship miniature dice whose side lengths have been cut in half. The dice are $\frac{1}{2}$ in. × $\frac{1}{2}$ in. × $\frac{1}{2}$ in. cubes. How many dice of this size can fit in the box?

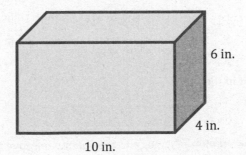

Example 2

A $\frac{1}{4}$ in. cube was used to fill the prism.

How many $\frac{1}{4}$ in. cubes does it take to fill the prism?

What is the volume of the prism?

How is the number of cubes related to the volume?

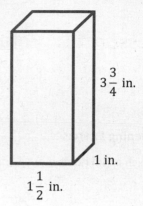

$3\frac{3}{4}$ in.

1 in.

$1\frac{1}{2}$ in.

Exercises

1. Use the prism to answer the following questions.

    a. Calculate the volume.

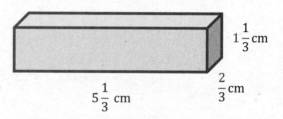

$1\frac{1}{3}$ cm

$\frac{2}{3}$ cm

$5\frac{1}{3}$ cm

    b. If you have to fill the prism with cubes whose side lengths are less than 1 cm, what size would be best?

    c. How many of the cubes would fit in the prism?

    d. Use the relationship between the number of cubes and the volume to prove that your volume calculation is correct.

EUREKA
MATH

2. Calculate the volume of the following rectangular prisms.

    a.

    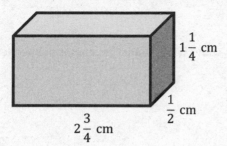

    $1\dfrac{1}{4}$ cm

    $\dfrac{1}{2}$ cm

    $2\dfrac{3}{4}$ cm

    b.

    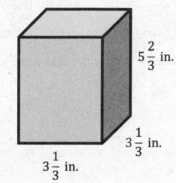

    $5\dfrac{2}{3}$ in.

    $3\dfrac{1}{3}$ in.

    $3\dfrac{1}{3}$ in.

3. A toy company is packaging its toys to be shipped. Each small toy is placed inside a cube-shaped box with side lengths of $\dfrac{1}{2}$ in. These smaller boxes are then placed into a larger box with dimensions of 12 in. × $4\dfrac{1}{2}$ in. × $3\dfrac{1}{2}$ in.

    a. What is the greatest number of small toy boxes that can be packed into the larger box for shipping?

    b. Use the number of small toy boxes that can be shipped in the larger box to help determine the volume of the shipping box.

4. A rectangular prism with a volume of 8 cubic units is filled with cubes twice: once with cubes with side lengths of $\frac{1}{2}$ unit and once with cubes with side lengths of $\frac{1}{3}$ unit.

    a. How many more of the cubes with $\frac{1}{3}$-unit side lengths than cubes with $\frac{1}{2}$-unit side lengths are needed to fill the prism?

    b. Why does it take more cubes with $\frac{1}{3}$-unit side lengths to fill the prism than it does with cubes with $\frac{1}{2}$-unit side lengths?

5. Calculate the volume of the rectangular prism. Show two different methods for determining the volume.

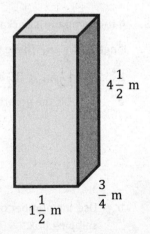

$4\frac{1}{2}$ m

$\frac{3}{4}$ m

$1\frac{1}{2}$ m

EUREKA
MATH

Problem Set

1. Answer the following questions using this rectangular prism:

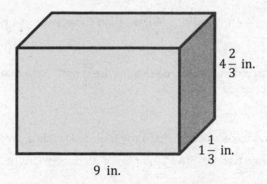

4 $\frac{2}{3}$ in.

1 $\frac{1}{3}$ in.

9 in.

   a.   What is the volume of the prism?

   b.   Linda fills the rectangular prism with cubes that have side lengths of $\frac{1}{3}$ in. How many cubes does she need to fill the rectangular prism?

   c.   How is the number of cubes related to the volume?

   d.   Why is the number of cubes needed different from the volume?

   e.   Should Linda try to fill this rectangular prism with cubes that are $\frac{1}{2}$ in. long on each side? Why or why not?

2. Calculate the volume of the following prisms.

   a.

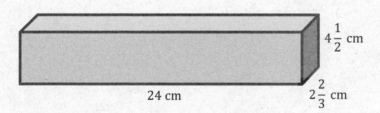

4 $\frac{1}{2}$ cm

24 cm

2 $\frac{2}{3}$ cm

   b.

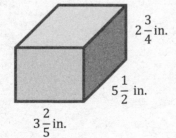

2 $\frac{3}{4}$ in.

5 $\frac{1}{2}$ in.

3 $\frac{2}{5}$ in.

3. A rectangular prism with a volume of 12 cubic units is filled with cubes twice: once with cubes with $\frac{1}{2}$-unit side lengths and once with cubes with $\frac{1}{3}$-unit side lengths.

    a. How many more of the cubes with $\frac{1}{3}$-unit side lengths than cubes with $\frac{1}{2}$-unit side lengths are needed to fill the prism?

    b. Finally, the prism is filled with cubes whose side lengths are $\frac{1}{4}$ unit. How many $\frac{1}{4}$-unit cubes would it take to fill the prism?

4. A toy company is packaging its toys to be shipped. Each toy is placed inside a cube-shaped box with side lengths of $3\frac{1}{2}$ in. These smaller boxes are then packed into a larger box with dimensions of 14 in. $\times$ 7 in. $\times$ $3\frac{1}{2}$ in.

    a. What is the greatest number of toy boxes that can be packed into the larger box for shipping?

    b. Use the number of toy boxes that can be shipped in the large box to determine the volume of the shipping box.

5. A rectangular prism has a volume of 34.224 cubic meters. The height of the box is 3.1 meters, and the length is 2.4 meters.

    a. Write an equation that relates the volume to the length, width, and height. Let w represent the width, in meters.

    b. Solve the equation.

EUREKA
MATH™

©2015 Great Minds eureka-math.org
G6-M5M6-SE-B3-1.3.1-01.2016

Lesson 12: From Unit Cubes to the Formulas for Volume

Classwork

Example 1

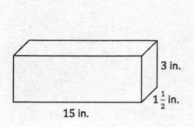 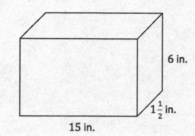 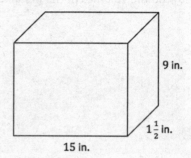

a.   Write a numerical expression for the volume of each of the rectangular prisms above.

b.   What do all of these expressions have in common? What do they represent?

c.   Rewrite the numerical expressions to show what they have in common.

d.   If we know volume for a rectangular prism as length times width times height, what is another formula for volume that we could use based on these examples?

e.   What is the area of the base for all of the rectangular prisms?

f.   Determine the volume of each rectangular prism using either method.

g.   How do the volumes of the first and second rectangular prisms compare? The volumes of the first and third?

Example 2

The base of a rectangular prism has an area of $3\frac{1}{4}$ in^2. The height of the prism is $2\frac{1}{2}$ in. Determine the volume of the rectangular prism.

Extension

A company is creating a rectangular prism that must have a volume of 6 ft^3. The company also knows that the area of the base must be $2\frac{1}{2}$ ft^2. How can you use what you learned today about volume to determine the height of the rectangular prism?

EUREKA
MATH™

©2015 Great Minds eureka-math.org
G6-M5M6-SE-B3-1.3.1-01.2016

Problem Set

1. Determine the volume of the rectangular prism.

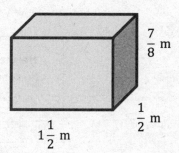

$\frac{7}{8}$ m

$\frac{1}{2}$ m

$1\frac{1}{2}$ m

2. The area of the base of a rectangular prism is $4\frac{3}{4}$ ft^2, and the height is $2\frac{1}{3}$ ft. Determine the volume of the rectangular prism.

3. The length of a rectangular prism is $3\frac{1}{2}$ times as long as the width. The height is $\frac{1}{4}$ of the width. The width is 3 cm. Determine the volume.

4.

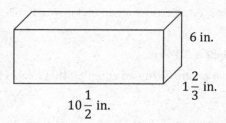

6 in.

$1\frac{2}{3}$ in.

$10\frac{1}{2}$ in.

   a. Write numerical expressions to represent the volume in two different ways, and explain what each reveals.

   b. Determine the volume of the rectangular prism.

5. An aquarium in the shape of a rectangular prism has the following dimensions: length = 50 cm, width = $25\frac{1}{2}$ cm, and height = $30\frac{1}{2}$ cm.

   a. Write numerical expressions to represent the volume in two different ways, and explain what each reveals.

   b. Determine the volume of the rectangular prism.

EUREKA
MATH™

6. The area of the base in this rectangular prism is fixed at 36 cm². As the height of the rectangular prism changes, the volume will also change as a result.

    a. Complete the table of values to determine the various heights and volumes.

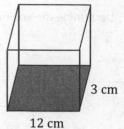

3 cm

12 cm

Height of Prism (in centimeters)	Volume of Prism (in cubic centimeters)
2	72
3	108
	144
	180
6	
7	
	288

    b. Write an equation to represent the relationship in the table. Be sure to define the variables used in the equation.

    c. What is the unit rate for this proportional relationship? What does it mean in this situation?

7. The volume of a rectangular prism is 16.328 cm³. The height is 3.14 cm.

    a. Let B represent the area of the base of the rectangular prism. Write an equation that relates the volume, the area of the base, and the height.

    b. Solve the equation for B.

Lesson 12:   From Unit Cubes to the Formulas for Volume

EUREKA MATH™

Lesson 13: The Formulas for Volume

Classwork

Example 1

Determine the volume of a cube with side lengths of $2\frac{1}{4}$ cm.

Example 2

Determine the volume of a rectangular prism with a base area of $\frac{7}{12}$ ft² and a height of $\frac{1}{3}$ ft.

Exercises

1. Use the rectangular prism to answer the next set of questions.
    a. Determine the volume of the prism.

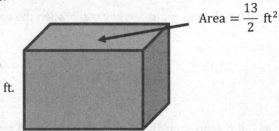

$\frac{5}{3}$ ft.

Area $= \frac{13}{2}$ ft²

    b. Determine the volume of the prism if the height of the prism is doubled.

c.   Compare the volume of the rectangular prism in part (a) with the volume of the prism in part (b). What do you notice?

d.   Complete and use the table below to determine the relationship between the height and volume.

Height of Prism (in feet)	Volume of Prism (in cubic feet)
$\dfrac{5}{3}$	$\dfrac{65}{6}$
$\dfrac{10}{3}$	$\dfrac{130}{6}$
$\dfrac{15}{3}$	
$\dfrac{20}{3}$	

What happened to the volume when the height was tripled?

What happened to the volume when the height was quadrupled?

What conclusions can you make when the base area stays constant and only the height changes?

2.

a.   If B represents the area of the base and h represents the height, write an expression that represents the volume.

EUREKA
MATH

     b.   If we double the height, write an expression for the new height.

     c.   Write an expression that represents the volume with the doubled height.

     d.   Write an equivalent expression using the commutative and associative properties to show the volume is twice the original volume.

3.   Use the cube to answer the following questions.

     a.   Determine the volume of the cube.

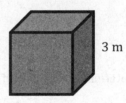

3 m

     b.   Determine the volume of a cube whose side lengths are half as long as the side lengths of the original cube.

     c.   Determine the volume if the side lengths are one-fourth as long as the original cube's side lengths.

     d.   Determine the volume if the side lengths are one-sixth as long as the original cube's side lengths.

©2015 Great Minds eureka-math.org
G6-M5M6-SE-B3-1.3.1-01.2016

e.   Explain the relationship between the side lengths and the volumes of the cubes.

4.   Check to see if the relationship you found in Exercise 3 is the same for rectangular prisms.

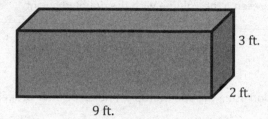

a.   Determine the volume of the rectangular prism.

b.   Determine the volume if all of the sides are half as long as the original lengths.

c.   Determine the volume if all of the sides are one-third as long as the original lengths.

d.   Is the relationship between the side lengths and the volume the same as the one that occurred in Exercise 3? Explain your answer.

©2015 Great Minds eureka-math.org
G6-M5M6-SE-B3-1.3.1-01.2016

5.

    a.   If e represents a side length of the cube, create an expression that shows the volume of the cube.

    b.   If we divide the side lengths by three, create an expression for the new side length.

    c.   Write an expression that represents the volume of the cube with one-third the side length.

    d.   Write an equivalent expression to show that the volume is $\frac{1}{27}$ of the original volume.

Problem Set

1. Determine the volume of the rectangular prism.

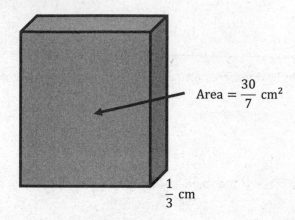

$$\text{Area} = \frac{30}{7} \text{ cm}^2$$

$\frac{1}{3}$ cm

2. Determine the volume of the rectangular prism in Problem 1 if the height is quadrupled (multiplied by four). Then, determine the relationship between the volumes in Problem 1 and this prism.

3. The area of the base of a rectangular prism can be represented by B, and the height is represented by h.

    a. Write an equation that represents the volume of the prism.

    b. If the area of the base is doubled, write an equation that represents the volume of the prism.

    c. If the height of the prism is doubled, write an equation that represents the volume of the prism.

    d. Compare the volume in parts (b) and (c). What do you notice about the volumes?

    e. Write an expression for the volume of the prism if both the height and the area of the base are doubled.

4. Determine the volume of a cube with a side length of $5\frac{1}{3}$ in.

5. Use the information in Problem 4 to answer the following:

    a. Determine the volume of the cube in Problem 4 if all of the side lengths are cut in half.

    b. How could you determine the volume of the cube with the side lengths cut in half using the volume in Problem 4?

EUREKA
MATH™

6. Use the rectangular prism to answer the following questions.

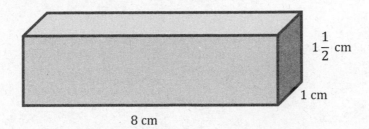

$1\frac{1}{2}$ cm

1 cm

8 cm

a. Complete the table.

Length of Prism	Volume of Prism
$l = 8$ cm	
$\frac{1}{2}l =$	
$\frac{1}{3}l =$	
$\frac{1}{4}l =$	
$2l =$	
$3l =$	
$4l =$	

b. How did the volume change when the length was one-third as long?

c. How did the volume change when the length was tripled?

d. What conclusion can you make about the relationship between the volume and the length?

7. The sum of the volumes of two rectangular prisms, Box A and Box B, are 14.325 cm³. Box A has a volume of 5.61 cm³.

a. Let B represent the volume of Box B in cubic centimeters. Write an equation that could be used to determine the volume of Box B.

b. Solve the equation to determine the volume of Box B.

c. If the area of the base of Box B is 1.5 cm², write an equation that could be used to determine the height of Box B. Let h represent the height of Box B in centimeters.

d. Solve the equation to determine the height of Box B.

This page intentionally left blank

Lesson 14: Volume in the Real World

Classwork

Example 1

a. The area of the base of a sandbox is $9\frac{1}{2}$ ft². The volume of the sandbox is $7\frac{1}{8}$ ft³. Determine the height of the sandbox.

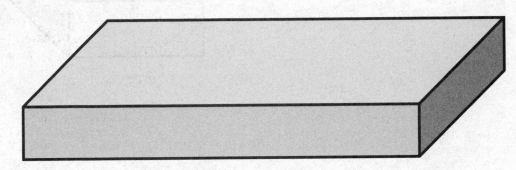

b. The sandbox was filled with sand, but after the kids played, some of the sand spilled out. Now, the sand is at a height of $\frac{1}{2}$ ft. Determine the volume of the sand in the sandbox after the children played in it.

Example 2

A special-order sandbox has been created for children to use as an archeological digging area at the zoo. Determine the volume of the sandbox.

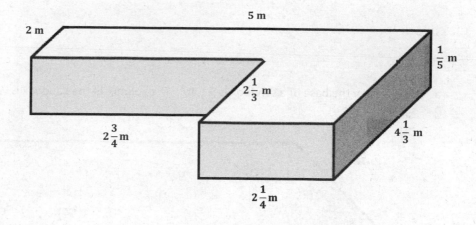

Exercises

1.

a. The volume of the rectangular prism is $\frac{35}{15}$ yd³. Determine the missing measurement using a one-step equation.

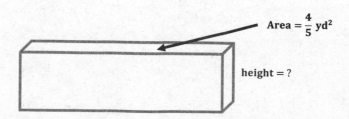

Area $= \frac{4}{5}$ yd²

height $= ?$

EUREKA
MATH™

b. The volume of the box is $\frac{45}{6}$ m³. Determine the area of the base using a one-step equation.

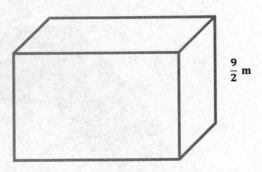

2. Marissa's fish tank needs to be filled with more water.

   a. Determine how much water the tank can hold.

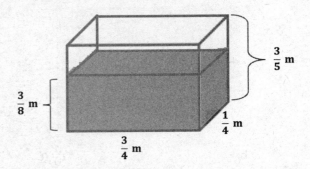

   b. Determine how much water is already in the tank.

   c. How much more water is needed to fill the tank?

3. Determine the volume of the composite figures.

    a.

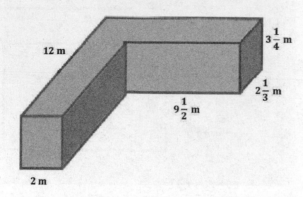

    b.

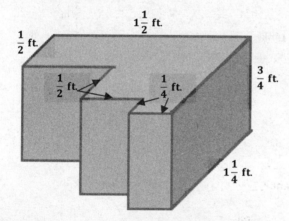

EUREKA
MATH™

Problem Set

1. The volume of a rectangular prism is $\frac{21}{12}$ ft³, and the height of the prism is $\frac{3}{4}$ ft. Determine the area of the base.

2. The volume of a rectangular prism is $\frac{10}{21}$ ft³. The area of the base is $\frac{2}{3}$ ft². Determine the height of the rectangular prism.

3. Determine the volume of the space in the tank that still needs to be filled with water if the water is $\frac{1}{3}$ ft. deep.

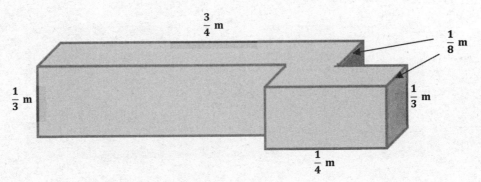

4. Determine the volume of the composite figure.

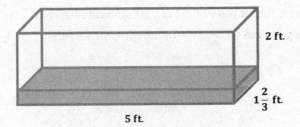

5. Determine the volume of the composite figure.

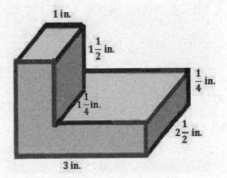

©2015 Great Minds eureka-math.org
G6-M5M6-SE-B3-1.3.1-01.2016

6.

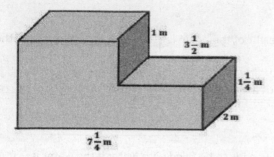

a.   Write an equation to represent the volume of the composite figure.

b.   Use your equation to calculate the volume of the composite figure.

©2015 Great Minds eureka-math.org
G6-M5M6-SE-B3-1.3.1-01.2016

Lesson 15: Representing Three-Dimensional Figures Using Nets

Classwork

Exercise: Cube

1. Nets are two-dimensional figures that can be folded into three-dimensional solids. Some of the drawings below are nets of a cube. Others are not cube nets; they can be folded, but not into a cube.

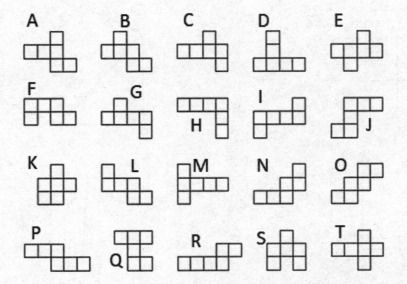

a. Experiment with the larger cut-out patterns provided. Shade in each of the figures above that can fold into a cube.

b. Write the letters of the figures that can be folded into a cube.

c. Write the letters of the figures that cannot be folded into a cube.

©2015 Great Minds eureka-math.org
G6-M5M6-SE-B3-1.3.1-01.2016

Lesson Summary

NET: If the surface of a 3-dimensional solid can be cut along sufficiently many edges so that the faces can be placed in one plane to form a connected figure, then the resulting system of faces is called a *net of the solid*.

Problem Set

1. Match the following nets to the picture of its solid. Then, write the name of the solid.

a.

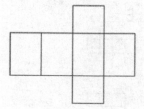

d.

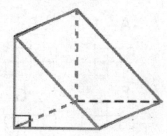

b.

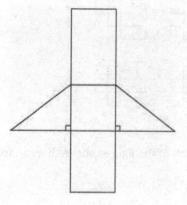

e.

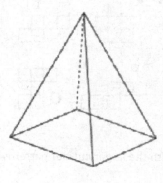

c.

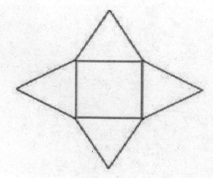

f.

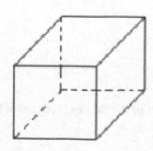

EUREKA
MATH

2. Sketch a net that can fold into a cube.

3. Below are the nets for a variety of prisms and pyramids. Classify the solids as prisms or pyramids, and identify the shape of the base(s). Then, write the name of the solid.

a.

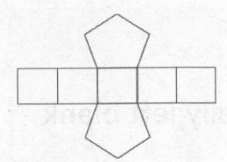

b.

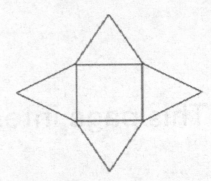

c.

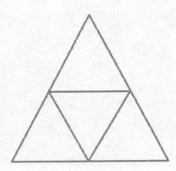

d.

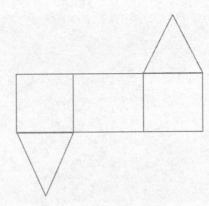

e.

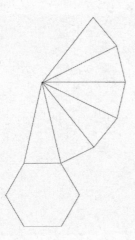

f.

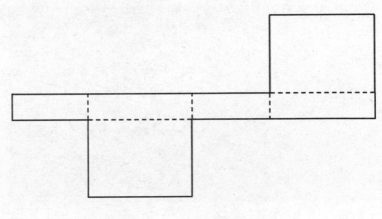

EUREKA
MATH™

This page intentionally left blank

Lesson 16: Constructing Nets

Classwork

Opening Exercise

Sketch the faces in the area below. Label the dimensions.

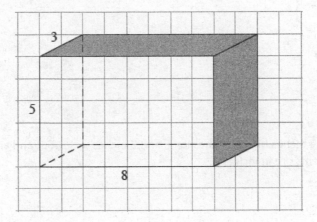

Exploratory Challenge 1: Rectangular Prisms

   a.   Use the measurements from the solid figures to cut and arrange the faces into a net. (Note: All measurements are in centimeters.)

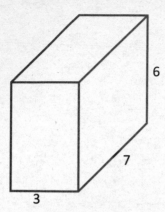

   b.   A juice box measures 4 inches high, 3 inches long, and 2 inches wide. Cut and arrange all 6 faces into a net. (Note: All measurements are in inches.)

   c.   Challenge: Write a numerical expression for the total area of the net for part (b). Explain each term in your expression.

Exploratory Challenge 2: Triangular Prisms

Use the measurements from the triangular prism to cut and arrange the faces into a net. (Note: All measurements are in inches.)

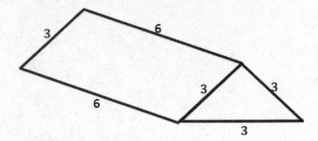

Exploratory Challenge 3: Pyramids

Pyramids are named for the shape of the base.

a.   Use the measurements from this square pyramid to cut and arrange the faces into a net. Test your net to be sure it folds into a square pyramid.

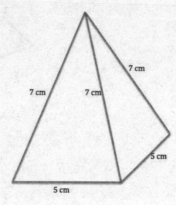

b.   A triangular pyramid that has equilateral triangles for faces is called a tetrahedron. Use the measurements from this tetrahedron to cut and arrange the faces into a net.

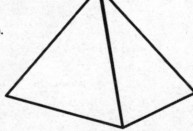

All edges are 4 in. in length.

EUREKA
MATH™

Problem Set

1. Sketch and label the net of the following solid figures, and label the edge lengths.

    a. A cereal box that measures 13 inches high, 7 inches long, and 2 inches wide

    b. A cubic gift box that measures 8 cm on each edge

    c. Challenge: Write a numerical expression for the total area of the net in part (b). Tell what each of the terms in your expression means.

2. This tent is shaped like a triangular prism. It has equilateral bases that measure 5 feet on each side. The tent is 8 feet long. Sketch the net of the tent, and label the edge lengths.

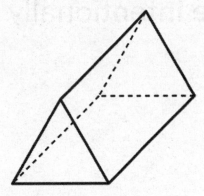

3. The base of a table is shaped like a square pyramid. The pyramid has equilateral faces that measure 25 inches on each side. The base is 25 inches long. Sketch the net of the table base, and label the edge lengths.

4. The roof of a shed is in the shape of a triangular prism. It has equilateral bases that measure 3 feet on each side. The length of the roof is 10 feet. Sketch the net of the roof, and label the edge lengths.

©2015 Great Minds eureka-math.org
G6-M5M6-SE-B3-1.3.1-01.2016

This page intentionally left blank

Lesson 17: From Nets to Surface Area

Classwork

Opening Exercise

 a. Write a numerical equation for the area of the figure below. Explain and identify different parts of the figure.

  i.

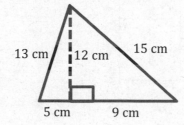

  ii. How would you write an equation that shows the area of a triangle with base b and height h?

 b. Write a numerical equation for the area of the figure below. Explain and identify different parts of the figure.

  i.

  ii. How would you write an equation that shows the area of a rectangle with base b and height h?

Example 1

Use the net to calculate the surface area of the figure. (Note: all measurements are in centimeters.)

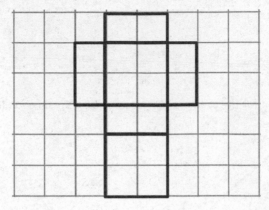

Example 2

Use the net to write an expression for surface area. (Note: all measurements are in square feet.)

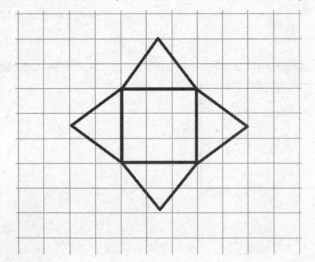

EUREKA
MATH™

Exercises

Name the solid the net would create, and then write an expression for the surface area. Use the expression to determine the surface area. Assume that each box on the grid paper represents a 1 cm × 1 cm square. Explain how the expression represents the figure.

1.

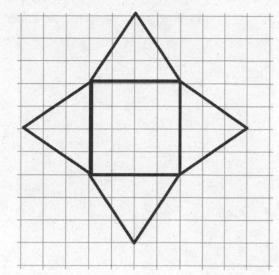

2.

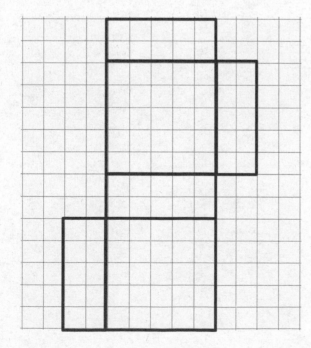

EUREKA MATH™

©2015 Great Minds eureka-math.org
G6-M5M6-SE-B3-1.3.1-01.2016

3.

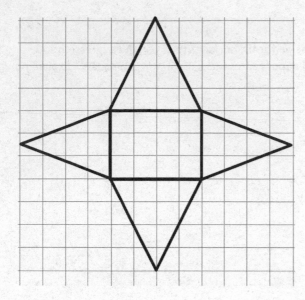

4.

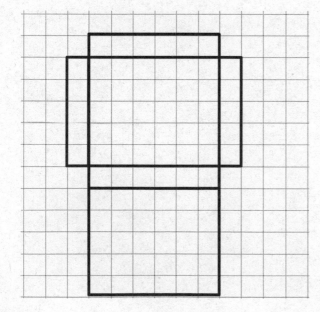

Lesson 17:   From Nets to Surface Area

©2015 Great Minds eureka-math.org
G6-M5M6-SE-B3-1.3.1-01.2016

Problem Set

Name the shape, and write an expression for surface area. Calculate the surface area of the figure. Assume each box on the grid paper represents a 1 ft. × 1 ft. square.

1.

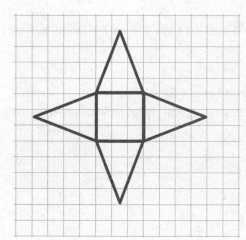

2.

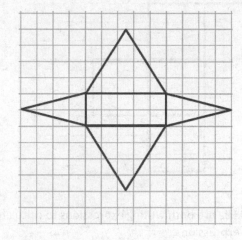

Explain the error in each problem below. Assume each box on the grid paper represents a 1 m × 1 m square.

3.

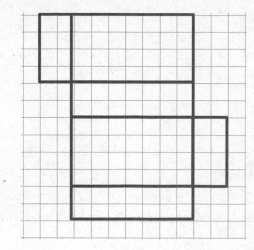

Name of Shape: Rectangular Pyramid, but more specifically a Square Pyramid

Area of Base: $3\text{ m} \times 3\text{ m} = 9\text{ m}^2$

Area of Triangles: $3\text{ m} \times 4\text{ m} = 12\text{ m}^2$

Surface Area: $9\text{ m}^2 + 12\text{ m}^2 + 12\text{ m}^2 + 12\text{ m}^2 + 12\text{ m}^2 = 57\text{ m}^2$

4.

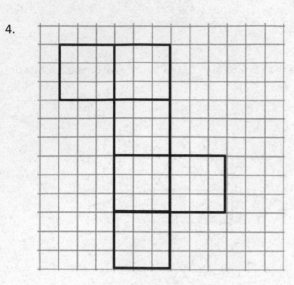

Name of Shape: Rectangular Prism or, more specifically, a Cube

Area of Faces: $3 \text{ m} \times 3 \text{ m} = 9 \text{ m}^2$

Surface Area: $9 \text{ m}^2 + 9 \text{ m}^2 + 9 \text{ m}^2 + 9 \text{ m}^2 + 9 \text{ m}^2 = 45 \text{ m}^2$

5. Sofia and Ella are both writing expressions to calculate the surface area of a rectangular prism. However, they wrote different expressions.

   a.   Examine the expressions below, and determine if they represent the same value. Explain why or why not.

Sofia's Expression:

$$(3 \text{ cm} \times 4 \text{ cm}) + (3 \text{ cm} \times 4 \text{ cm}) + (3 \text{ cm} \times 5 \text{ cm}) + (3 \text{ cm} \times 5 \text{ cm}) + (4 \text{ cm} \times 5 \text{ cm}) + (4 \text{ cm} \times 5 \text{ cm})$$

Ella's Expression:

$$2(3 \text{ cm} \times 4 \text{ cm}) + 2(3 \text{ cm} \times 5 \text{ cm}) + 2(4 \text{ cm} \times 5 \text{ cm})$$

   b.   What fact about the surface area of a rectangular prism does Ella's expression show more clearly than Sofia's?

EUREKA
MATH™

Lesson 18: Determining Surface Area of Three-Dimensional Figures

Classwork

Opening Exercise

a.   What three-dimensional figure does the net create?

b.   Measure (in inches) and label each side of the figure.

c.   Calculate the area of each face, and record this value inside the corresponding rectangle.

d.   How did we compute the surface area of solid figures in previous lessons?

e.   Write an expression to show how we can calculate the surface area of the figure above.

f.   What does each part of the expression represent?

g.   What is the surface area of the figure?

Example 1

Fold the net used in the Opening Exercise to make a rectangular prism. Have the two faces with the largest area be the bases of the prism. Fill in the first row of the table below.

Area of Top (base)	Area of Bottom (base)	Area of Front	Area of Back	Area of Left Side	Area of Right Side

Examine the rectangular prism below. Complete the table.

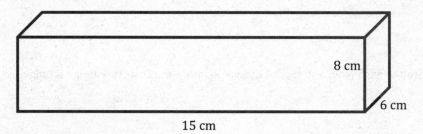

8 cm

6 cm

15 cm

Area of Top (base)	Area of Bottom (base)	Area of Front	Area of Back	Area of Left Side	Area of Right Side

EUREKA
MATH™

©2015 Great Minds eureka-math.org
G6-M5M6-SE-B3-1.3.1-01.2016

Example 2

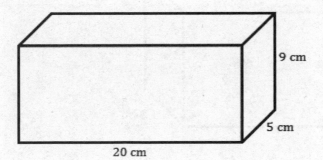

Exercises 1–3

1. Calculate the surface area of each of the rectangular prisms below.

a.

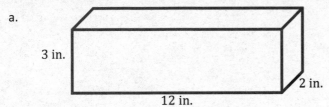

b.

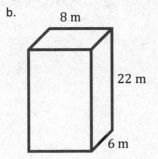

c.

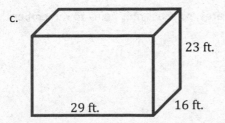

d.

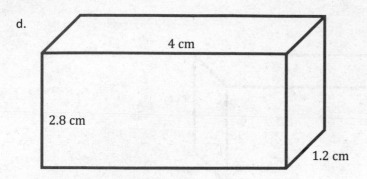

2. Calculate the surface area of the cube.

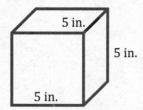

3. All the edges of a cube have the same length. Tony claims that the formula $SA = 6s^2$, where s is the length of each side of the cube, can be used to calculate the surface area of a cube.

a. Use the dimensions from the cube in Problem 2 to determine if Tony's formula is correct.

b. Why does this formula work for cubes?

c. Becca does not want to try to remember two formulas for surface area, so she is only going to remember the formula for a cube. Is this a good idea? Why or why not?

©2015 Great Minds eureka-math.org
G6-M5M6-SE-B3-1.3.1-01.2016

EUREKA
MATH™

Lesson Summary

Surface Area Formula for a Rectangular Prism: $SA = 2lw + 2lh + 2wh$

Surface Area Formula for a Cube: $SA = 6s^2$

Problem Set

Calculate the surface area of each figure below. Figures are not drawn to scale.

1.

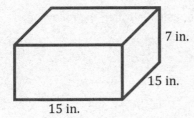

7 in.

15 in.

15 in.

2.

2.3 cm

8.4 cm

18.7 cm

3.

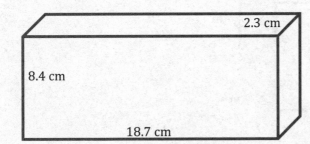

$2\frac{1}{3}$ ft.

$2\frac{1}{3}$ ft.

$2\frac{1}{3}$ ft.

4.

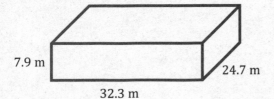

7.9 m

24.7 m

32.3 m

EUREKA
MATH™

5. Write a numerical expression to show how to calculate the surface area of the rectangular prism. Explain each part of the expression.

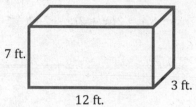

7 ft.

12 ft.

3 ft.

6. When Louie was calculating the surface area for Problem 4, he identified the following:

    length = 24.7 m, width = 32.3 m, and height = 7.9 m.

    However, when Rocko was calculating the surface area for the same problem, he identified the following:

    length = 32.3 m, width = 24.7 m, and height = 7.9 m.

    Would Louie and Rocko get the same answer? Why or why not?

7. Examine the figure below.

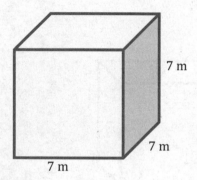

7 m

7 m

7 m

    a.   What is the most specific name of the three-dimensional shape?

    b.   Write two different expressions for the surface area.

    c.   Explain how these two expressions are equivalent.

EUREKA
MATH

Lesson 19: Surface Area and Volume in the Real World

Classwork

Opening Exercise

A box needs to be painted. How many square inches need to be painted to cover the entire surface of the box?

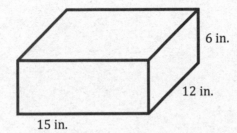

A juice box is 4 in. tall, 1 in. wide, and 2 in. long. How much juice fits inside the juice box?

How did you decide how to solve each problem?

Discussion

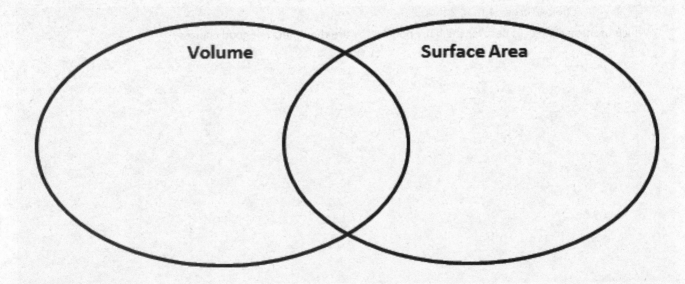

Example 1

Vincent put logs in the shape of a rectangular prism outside his house. However, it is supposed to snow, and Vincent wants to buy a cover so the logs stay dry. If the pile of logs creates a rectangular prism with these measurements:

33 cm long, 12 cm wide, and 48 cm high,

what is the minimum amount of material needed to cover the pile of logs?

Exercises

Use your knowledge of volume and surface area to answer each problem.

1. Quincy Place wants to add a pool to the neighborhood. When determining the budget, Quincy Place determined that it would also be able to install a baby pool that required less than 15 cubic feet of water. Quincy Place has three different models of a baby pool to choose from.

        Choice One: 5 ft. × 5 ft. × 1 ft.

        Choice Two: 4 ft. × 3 ft. × 1 ft.

        Choice Three: 4 ft. × 2 ft. × 2 ft.

Which of these choices is best for the baby pool? Why are the others not good choices?

©2015 Great Minds eureka-math.org
G6-M5M6-SE-B3-1.3.1-01.2016

2. A packaging firm has been hired to create a box for baby blocks. The firm was hired because it could save money by creating a box using the least amount of material. The packaging firm knows that the volume of the box must be 18 cm^3.

   a. What are possible dimensions for the box if the volume must be exactly 18 cm^3?

   b. Which set of dimensions should the packaging firm choose in order to use the least amount of material? Explain.

3. A gift has the dimensions of $50 \text{ cm} \times 35 \text{ cm} \times 5 \text{ cm}$. You have wrapping paper with dimensions of $75 \text{ cm} \times 60 \text{ cm}$. Do you have enough wrapping paper to wrap the gift? Why or why not?

4. Tony bought a flat-rate box from the post office to send a gift to his mother for Mother's Day. The dimensions of the medium-size box are $14 \text{ inches} \times 12 \text{ inches} \times 3.5 \text{ inches}$. What is the volume of the largest gift he can send to his mother?

©2015 Great Minds eureka-math.org
G6-M5M6-SE-B3-1.3.1-01.2016

5. A cereal company wants to change the shape of its cereal box in order to attract the attention of shoppers. The original cereal box has dimensions of 8 inches × 3 inches × 11 inches. The new box the cereal company is thinking of would have dimensions of 10 inches × 10 inches × 3 inches.

    a. Which box holds more cereal?

    b. Which box requires more material to make?

6. Cinema theaters created a new popcorn box in the shape of a rectangular prism. The new popcorn box has a length of 6 inches, a width of 3.5 inches, and a height of 3.5 inches but does not include a lid.

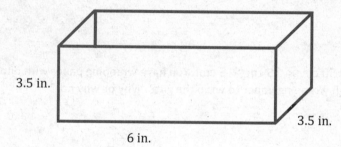

3.5 in.

6 in.

3.5 in.

    a. How much material is needed to create the box?

    b. How much popcorn does the box hold?

Problem Set

Solve each problem below.

1. Dante built a wooden, cubic toy box for his son. Each side of the box measures 2 feet.

    a.   How many square feet of wood did he use to build the box?

    b.   How many cubic feet of toys will the box hold?

2. A company that manufactures gift boxes wants to know how many different-sized boxes having a volume of 50 cubic centimeters it can make if the dimensions must be whole centimeters.

    a.   List all the possible whole number dimensions for the box.

    b.   Which possibility requires the least amount of material to make?

    c.   Which box would you recommend the company use? Why?

3. A rectangular box of rice is shown below. What is the greatest amount of rice, in cubic inches, that the box can hold?

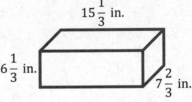

4. The Mars Cereal Company has two different cereal boxes for Mars Cereal. The large box is 8 inches wide, 11 inches high, and 3 inches deep. The small box is 6 inches wide, 10 inches high, and 2.5 inches deep.

    a.   How much more cardboard is needed to make the large box than the small box?

    b.   How much more cereal does the large box hold than the small box?

5. A swimming pool is 8 meters long, 6 meters wide, and 2 meters deep. The water-resistant paint needed for the pool costs $6 per square meter. How much will it cost to paint the pool?

    a.   How many faces of the pool do you have to paint?

    b.   How much paint (in square meters) do you need to paint the pool?

    c.   How much will it cost to paint the pool?

6. Sam is in charge of filling a rectangular hole with cement. The hole is 9 feet long, 3 feet wide, and 2 feet deep. How much cement will Sam need?

EUREKA
MATH™

Lesson 19:   Surface Area and Volume in the Real World

S.111

©2015 Great Minds eureka-math.org
G6-M5M6-SE-B3-1.3.1-01.2016

7. The volume of Box D subtracted from the volume of Box C is 23.14 cubic centimeters. Box D has a volume of 10.115 cubic centimeters.

    a. Let C be the volume of Box C in cubic centimeters. Write an equation that could be used to determine the volume of Box C.

    b. Solve the equation to determine the volume of Box C.

    c. The volume of Box C is one-tenth the volume another box, Box E. Let E represent the volume of Box E in cubic centimeters. Write an equation that could be used to determine the volume of Box E, using the result from part (b).

    d. Solve the equation to determine the volume of Box E.

©2015 Great Minds eureka-math.org
G6-M5M6-SE-B3-1.3.1-01.2016

Lesson 19a: Applying Surface Area and Volume to Aquariums

Classwork

Opening Exercise

Determine the volume of this aquarium.

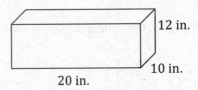

12 in.

10 in.

20 in.

Mathematical Modeling Exercise: Using Ratios and Unit Rate to Determine Volume

For his environmental science project, Jamie is creating habitats for various wildlife including fish, aquatic turtles, and aquatic frogs. For each of these habitats, he uses a standard aquarium with length, width, and height dimensions measured in inches, identical to the aquarium mentioned in the Opening Exercise. To begin his project, Jamie needs to determine the volume, or cubic inches, of water that can fill the aquarium.

Use the table below to determine the unit rate of gallons/cubic inches.

Gallons	Cubic Inches
1	
2	462
3	693
4	924
5	1,155

Determine the volume of the aquarium.

©2015 Great Minds eureka-math.org
G6-M5M6-SE-B3-1.3.1-01.2016

Exercise 1

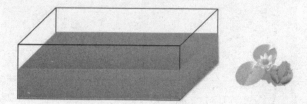

a.   Determine the volume of the tank when filled with 7 gallons of water.

b.   Work with your group to determine the height of the water when Jamie places 7 gallons of water in the aquarium.

Exercise 2

a.   Use the table from Example 1 to determine the volume of the aquarium when Jamie pours 3 gallons of water into the tank.

b.   Use the volume formula to determine the missing height dimension.

©2015 Great Minds eureka-math.org
G6-M5M6-SE-B3-1.3.1-01.2016

Exercise 3

    a.   Using the table of values below, determine the unit rate of liters to gallon.

Gallons	Liters
1	
2	7.57
4	15.14

    b.   Using this conversion, determine the number of liters needed to fill the 10-gallon tank.

    c.   The ratio of the number of centimeters to the number of inches is 2.54: 1. What is the unit rate?

    d.   Using this information, complete the table to convert the heights of the water in inches to the heights of the water in centimeters Jamie will need for his project at home.

Height (in inches)	Convert to Centimeters	Height (in centimeters)
1	$2.54 \dfrac{\text{centimeters}}{\text{inch}} \times 1 \text{ inch}$	2.54
3.465		
8.085		
11.55		

Exercise 4

a. Determine the amount of plastic film the manufacturer uses to cover the aquarium faces. Draw a sketch of the aquarium to assist in your calculations. Remember that the actual height of the aquarium is 12 inches.

b. We do not include the measurement of the top of the aquarium since it is open without glass and does not need to be covered with film. Determine the area of the top of the aquarium, and find the amount of film the manufacturer uses to cover only the sides, front, back, and bottom.

c. Since Jamie needs three aquariums, determine the total surface area of the three aquariums.

©2015 Great Minds eureka-math.org
G6-M5M6-SE-B3-1.3.1-01.2016

EUREKA
MATH

Problem Set

This Problem Set is a culmination of skills learned in this module. Note that the figures are not drawn to scale.

1. Calculate the area of the figure below.

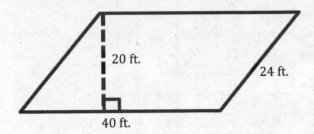

2. Calculate the area of the figure below.

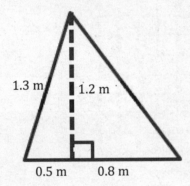

3. Calculate the area of the figure below.

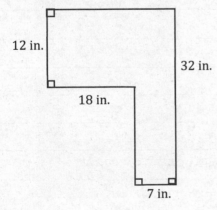

EUREKA
MATH™

4. Complete the table using the diagram on the coordinate plane.

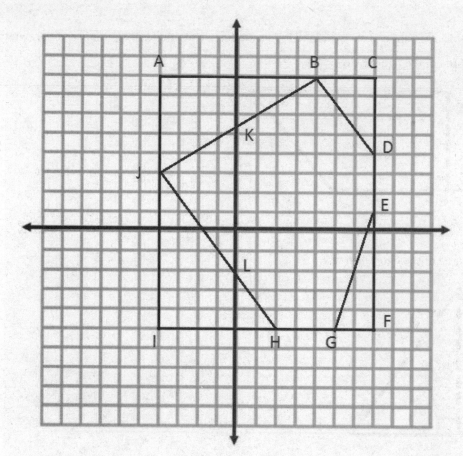

Line Segment	Point	Point	Distance	Proof
$\overline{AB}$				
$\overline{CE}$				
$\overline{GI}$				
$\overline{HI}$				
$\overline{IJ}$				
$\overline{AI}$				
$\overline{AJ}$				

EUREKA MATH™

©2015 Great Minds eureka-math.org
G6-M5M6-SE-B3-1.3.1-01.2016

5. Plot the points below, and draw the shape. Then, determine the area of the polygon.

$A(-3, 5), B(4, 3), C(0, -5)$

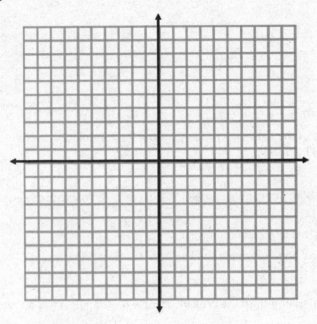

6. Determine the volume of the figure.

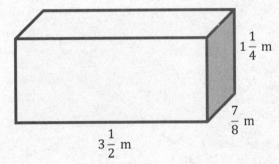

$1\frac{1}{4}$ m

$\frac{7}{8}$ m

$3\frac{1}{2}$ m

7. Give at least three more expressions that could be used to determine the volume of the figure in Problem 6.

8. Determine the volume of the irregular figure.

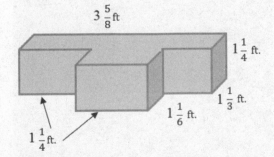

$3\frac{5}{8}$ ft

$1\frac{1}{4}$ ft.

$1\frac{1}{3}$ ft.

$1\frac{1}{6}$ ft.

$1\frac{1}{4}$ ft.

9. Draw and label a net for the following figure. Then, use the net to determine the surface area of the figure.

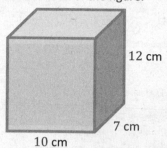

12 cm

7 cm

10 cm

10. Determine the surface area of the figure in Problem 9 using the formula $SA = 2lw + 2lh + 2wh$. Then, compare your answer to the solution in Problem 9.

11. A parallelogram has a base of 4.5 cm and an area of 9.495 cm². Tania wrote the equation $4.5x = 9.495$ to represent this situation.

    a. Explain what x represents in the equation.

    b. Solve the equation for x and determine the height of the parallelogram.

12. Triangle A has an area equal to one-third the area of Triangle B. Triangle A has an area of $3\frac{1}{2}$ square meters.

    a. Gerard wrote the equation $\frac{B}{3} = 3\frac{1}{2}$. Explain what B represents in the equation.

    b. Determine the area of Triangle B.

EUREKA
MATH™

©2015 Great Minds eureka-math.org
G6-M5M6-SE-B3-1.3.1-01.2016

Eureka Math
Grade 6
Module 6

Special thanks go to the Gordan A. Cain Center and to the Department of Mathematics at Louisiana State University for their support in the development of *Eureka Math*.

Printed in the U.S.A.

This book may be purchased from the publisher at eureka-math.org

10 9 8 7 6 5 4 3

ISBN 978-1-63255-314-0

Lesson 1: Posing Statistical Questions

Classwork

Example 1: Using Data to Answer Questions

Honeybees are important because they produce honey and pollinate plants. Since 2007, there has been a decline in the honeybee population in the United States. Honeybees live in hives, and a beekeeper in Wisconsin notices that this year, he has 5 fewer hives of bees than last year. He wonders if other beekeepers in Wisconsin are also losing hives. He decides to survey other beekeepers and ask them if they have fewer hives this year than last year, and if so, how many fewer. He then uses the data to conclude that most beekeepers have fewer hives this year than last and that a typical decrease is about 4 hives.

Statistics is about using data to answer questions. In this module, you will use the following four steps in your work with data:

    Step 1: Pose a question that can be answered by data.

    Step 2: Determine a plan to collect the data.

    Step 3: Summarize the data with graphs and numerical summaries.

    Step 4: Answer the question posed in Step 1 using the data and summaries.

You will be guided through this process as you study these lessons. This first lesson is about the first step: What is a statistical question, and what does it mean that a question can be answered by data?

Example 2: What Is a Statistical Question?

Jerome, a sixth grader at Roosevelt Middle School, is a huge baseball fan. He loves to collect baseball cards. He has cards of current players and of players from past baseball seasons. With his teacher's permission, Jerome brought his baseball card collection to school. Each card has a picture of a current or past major league baseball player, along with information about the player. When he placed his cards out for the other students to see, they asked Jerome all sorts of questions about his cards. Some asked:

- What is Jerome's favorite card?
- What is the typical cost of a card in Jerome's collection? For example, what is the average cost of a card?
- Are more of Jerome's cards for current players or for past players?
- Which card is the newest card in Jerome's collection?

©2015 Great Minds eureka-math.org
G6-M5M6-SE-B3-1.3.1-01.2016

Exercises 1–5

1. For each of the following, determine whether or not the question is a statistical question. Give a reason for your answer.

   a. Who is my favorite movie star?

   b. What are the favorite colors of sixth graders in my school?

   c. How many years have students in my school's band or orchestra played an instrument?

   d. What is the favorite subject of sixth graders at my school?

   e. How many brothers and sisters does my best friend have?

2. Explain why each of the following questions is not a statistical question.

   a. How old am I?

   b. What's my favorite color?

   c. How old is the principal at our school?

3. Ronnie, a sixth grader, wanted to find out if he lived the farthest from school. Write a statistical question that would help Ronnie find the answer.

4. Write a statistical question that can be answered by collecting data from students in your class.

5. Change the following question to make it a statistical question: How old is my math teacher?

Example 3: Types of Data

We use two types of data to answer statistical questions: numerical data and categorical data. If you recorded the ages of 25 baseball cards, we would have numerical data. Each value in a numerical data set is a number. If we recorded the team of the featured player for each of 25 baseball cards, you would have categorical data. Although you still have 25 data values, the data values are not numbers. They would be team names, which you can think of as categories.

Exercises 6–7

6. Identify each of the following data sets as categorical (C) or numerical (N).

    a. Heights of 20 sixth graders _____

    b. Favorite flavor of ice cream for each of 10 sixth graders _____

    c. Hours of sleep on a school night for each of 30 sixth graders _____

    d. Type of beverage drunk at lunch for each of 15 sixth graders _____

    e. Eye color for each of 30 sixth graders _____

    f. Number of pencils in the desk of each of 15 sixth graders _____

7. For each of the following statistical questions, identify whether the data Jerome would collect to answer the question would be numerical or categorical. Explain your answer, and list four possible data values.

   a.   How old are the cards in the collection?

   b.   How much did the cards in the collection cost?

   c.   Where did Jerome get the cards in the collection?

©2015 Great Minds eureka-math.org
G6-M5M6-SE-B3-1.3.1-01.2016

Problem Set

1. For each of the following, determine whether the question is a statistical question. Give a reason for your answer.
   a. How many letters are in my last name?
   b. How many letters are in the last names of the students in my sixth-grade class?
   c. What are the colors of the shoes worn by students in my school?
   d. What is the maximum number of feet that roller coasters drop during a ride?
   e. What are the heart rates of students in a sixth-grade class?
   f. How many hours of sleep per night do sixth graders usually get when they have school the next day?
   g. How many miles per gallon do compact cars get?

2. Identify each of the following data sets as categorical (C) or numerical (N). Explain your answer.
   a. Arm spans of 12 sixth graders
   b. Number of languages spoken by each of 20 adults
   c. Favorite sport of each person in a group of 20 adults
   d. Number of pets for each of 40 third graders
   e. Number of hours a week spent reading a book for a group of middle school students

3. Rewrite each of the following questions as a statistical question.
   a. How many pets does your teacher have?
   b. How many points did the high school soccer team score in its last game?
   c. How many pages are in our math book?
   d. Can I do a handstand?

4.   Write a statistical question that would be answered by collecting data from the sixth graders in your classroom.

5.   Are the data you would collect to answer the question you wrote in Problem 4 categorical or numerical? Explain your answer.

Lesson 2: Displaying a Data Distribution

Classwork

Example 1: Heart Rate

Mia, a sixth grader at Roosevelt Middle School, was thinking about joining the middle school track team. She read that Olympic athletes have lower resting heart rates than most people. She wondered about her own heart rate and how it would compare to other students. Mia was interested in investigating the statistical question: What are the heart rates of students in my sixth-grade class?

Heart rates are expressed as beats per minute (or bpm). Mia knew her resting heart rate was 80 beats per minute. She asked her teacher if she could collect the heart rates of the other students in her class. With the teacher's help, the other sixth graders in her class found their heart rates and reported them to Mia. The following numbers are the resting heart rates (in beats per minute) for the 22 other students in Mia's class.

89 87 85 84 90 79 83 85 86 88 84 81 88 85 83 83 86 82 83 86 82 84

Exercises 1–10

1. What was the heart rate for the student with the lowest heart rate?

2. What was the heart rate for the student with the highest heart rate?

3. How many students had a heart rate greater than 86 bpm?

4. What fraction of students had a heart rate less than 82 bpm?

5. What heart rate occurred most often?

6. What heart rate describes the center of the data?

7. Some students had heart rates that were unusual in that they were quite a bit higher or quite a bit lower than most other students' heart rates. What heart rates would you consider unusual?

8. If Mia's teacher asked what the typical heart rate is for sixth graders in the class, what would you tell Mia's teacher?

9. Remember that Mia's heart rate was 80 bpm. Add a dot for Mia's heart rate to the dot plot in Example 1.

10. How does Mia's heart rate compare with the heart rates of the other students in the class?

Example 2: Seeing the Spread in Dot Plots

Mia's class collected data to answer several other questions about her class. After collecting the data, they drew dot plots of their findings.

One student collected data to answer the question: How many textbooks are in the desks or lockers of sixth graders? She made the following dot plot, not including her data.

Dot Plot of Number of Textbooks

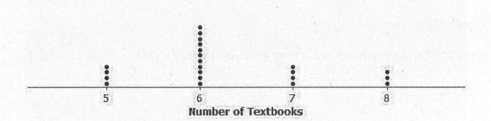

Another student in Mia's class wanted to ask the question: How tall are the sixth graders in our class?

This dot plot shows the heights of the sixth graders in Mia's class, not including the datum for the student conducting the survey.

Dot Plot of Height

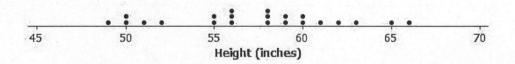

Exercises 11–14

Below are four statistical questions and four different dot plots of data collected to answer these questions. Match each statistical question with the appropriate dot plot, and explain each choice.

Statistical Questions:

11. What are the ages of fourth graders in our school?

12. What are the heights of the players on the eighth-grade boys' basketball team?

13. How many hours of TV do sixth graders in our class watch on a school night?

14. How many different languages do students in our class speak?

Dot Plot A   **Dot Plot B**

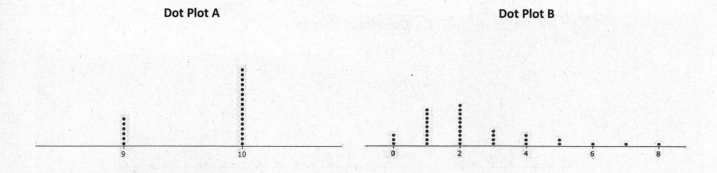

Dot Plot C   **Dot Plot D**

EUREKA
MATH

Problem Set

1. The dot plot below shows the vertical jump height (in inches) of some NBA players. A vertical jump height is how high a player can jump from a standstill.

Dot Plot of Vertical Jump

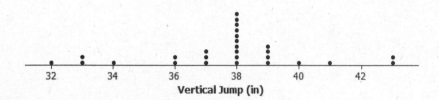

Vertical Jump (in)

   a. What statistical question do you think could be answered using these data?

   b. What was the highest vertical jump by a player?

   c. What was the lowest vertical jump by a player?

   d. What is the most common vertical jump height (the height that occurred most often)?

   e. How many players jumped the most common vertical jump height?

   f. How many players jumped higher than 40 inches?

   g. Another NBA player jumped 33 inches. Add a dot for this player on the dot plot. How does this player compare with the other players?

2. Below are two statistical questions and two different dot plots of data collected to answer these questions. Match each statistical question with its dot plot, and explain each choice.

   Statistical Questions:

   a. What is the number of fish (if any) that students in class have in an aquarium at their homes?

   b. How many days out of the week do the children on my street go to the playground?

Dot Plot A   **Dot Plot B**

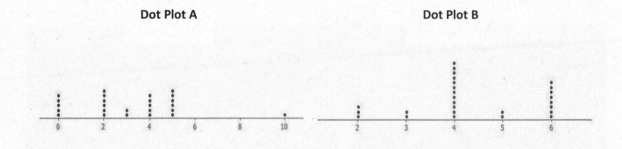

3. Read each of the following statistical questions. Write a description of what the dot plot of data collected to answer the question might look like. Your description should include a description of the spread of the data and the center of the data.

    a. What is the number of hours sixth graders are in school during a typical school day?

    b. What is the number of video games owned by the sixth graders in our class?

Lesson 3: Creating a Dot Plot

Classwork

Example 1: Hours of Sleep

Robert, a sixth grader at Roosevelt Middle School, usually goes to bed around 10:00 p.m. and gets up around 6:00 a.m. to get ready for school. That means he gets about 8 hours of sleep on a school night. He decided to investigate the statistical question: How many hours per night do sixth graders usually sleep when they have school the next day?

Robert took a survey of 29 sixth graders and collected the following data to answer the question.

7 8 5 9 9 9 7 7 10 10 11 9 8 8 8 12 6 11 10 8 8 9 9 9 8 10 9 9 8

Robert decided to make a dot plot of the data to help him answer his statistical question. Robert first drew a number line and labeled it from 5 to 12 to match the lowest and highest number of hours slept. Robert's datum is not included.

Dot Plot of Number of Hours Slept

5   6   7   8   9   10   11   12
Number of Hours Slept

He then placed a dot above 7 for the first value in the data set. He continued to place dots above the numbers until each number in the data set was represented by a dot.

Dot Plot of Number of Hours Slept

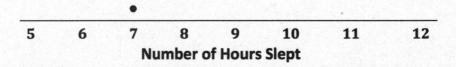

5   6   7   8   9   10   11   12
Number of Hours Slept

©2015 Great Minds eureka-math.org
G6-M5M6-SE-B3-1.3.1-01.2016

Exercises 1–9

1. Complete Robert's dot plot by placing a dot above the corresponding number on the number line for each value in the data set. If there is already a dot above a number, then add another dot above the dot already there. Robert's datum is not included.

2. What are the least and the most hours of sleep reported in the survey of sixth graders?

3. What number of hours slept occurred most often in the data set?

4. What number of hours of sleep would you use to describe the center of the data?

5. Think about how many hours of sleep you usually get on a school night. How does your number compare with the number of hours of sleep from the survey of sixth graders?

Here are the data for the number of hours the sixth graders usually sleep when they do not have school the next day.

     7 8 10 11 5 6 12 13 13 7 9 8 10 12 11 12 8 9 10 11 10 12 11 11 11 12 11 11 10

6. Make a dot plot of the number of hours slept when there is no school the next day.

7. When there is no school the next day, what number of hours of sleep would you use to describe the center of the data?

8. What are the least and most number of hours slept with no school the next day reported in the survey?

9. Do students tend to sleep longer when they do not have school the next day than when they do have school the next day? Explain your answer using the data in both dot plots.

Example 2: Building and Interpreting a Frequency Table

A group of sixth graders investigated the statistical question, "How many hours per week do sixth graders spend playing a sport or an outdoor game?"

Here are the data students collected from a sample of 26 sixth graders showing the number of hours per week spent playing a sport or a game outdoors.

3 2 0 6 3 3 3 1 1 2 2 8 12 4 4 4 3 3 1 1 0 0 6 2 3 2

To help organize the data, students summarized the data in a frequency table. A frequency table lists possible data values and how often each value occurs.

To build a frequency table, first make three columns. Label one column "Number of Hours Playing a Sport/Game," label the second column "Tally," and label the third column "Frequency." Since the least number of hours was 0 and the most was 12, list the numbers from 0 to 12 in the "Number of Hours" column.

Exercises 10–15

10. Complete the tally mark column in the table created in Example 2.

11. For each number of hours, find the total number of tally marks, and place this in the frequency column in the table created in Example 2.

12. Make a dot plot of the number of hours playing a sport or playing outdoors.

13. What number of hours describes the center of the data?

14. How many of the sixth graders reported that they spend eight or more hours a week playing a sport or playing outdoors?

15. The sixth graders wanted to answer the question, "How many hours do sixth graders spend per week playing a sport or playing an outdoor game?" Using the frequency table and the dot plot, how would you answer the sixth graders' question?

Problem Set

1. The data below are the number of goals scored by a professional indoor soccer team over its last 23 games.

    8 16 10 9 11 11 10 15 16 11 15 13 8 9 11 9 8 11 16 15 10 9 12

    a. Make a dot plot of the number of goals scored.

    b. What number of goals describes the center of the data?

    c. What is the least and most number of goals scored by the team?

    d. Over the 23 games played, the team lost 10 games. Circle the dots on the plot that you think represent the games that the team lost. Explain your answer.

2. A sixth grader rolled two number cubes 21 times. The student found the sum of the two numbers that he rolled each time. The following are the sums for the 21 rolls of the two number cubes.

    9 2 4 6 5 7 8 11 9 4 6 5 7 7 8 8 7 5 7 6 6

    a. Complete the frequency table.

Sum Rolled	Tally	Frequency
2		
3		
4		
5		
6		
7		
8		
9		
10		
11		
12		

    b. What sum describes the center of the data?

    c. What sum occurred most often for these 21 rolls of the number cubes?

©2015 Great Minds eureka-math.org
G6-M5M6-SE-B3-1.3.1-01.2016

3. The dot plot below shows the number of raisins in 25 small boxes of raisins.

Dot Plot of Number of Raisins

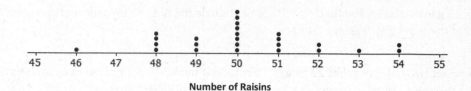

Number of Raisins

a. Complete the frequency table.

Number of Raisins	Tally	Frequency
46		
47		
48		
49		
50		
51		
52		
53		
54		

b. Another student opened up a box of raisins and reported that it had 63 raisins. Do you think that this student had the same size box of raisins? Why or why not?

EUREKA
MATH™

©2015 Great Minds eureka-math.org
G6-M5M6-SE-B3-1.3.1-01.2016

Lesson 4: Creating a Histogram

Classwork

Example 1: Frequency Table with Intervals

The boys' and girls' basketball teams at Roosevelt Middle School wanted to raise money to help buy new uniforms. They decided to sell baseball caps with the school logo on the front to family members and other interested fans. To obtain the correct cap size, students had to measure the head circumference (distance around the head) of the adults who wanted to order a cap. The following data set represents the head circumferences, in millimeters (mm), of the adults.

513, 525, 531, 533, 535, 535, 542, 543, 546, 549, 551, 552, 552, 553, 554, 555, 560, 561, 563, 563, 563, 565,

565, 568, 568, 571, 571, 574, 577, 580, 583, 583, 584, 585, 591, 595, 598, 603, 612, 618

The caps come in six sizes: XS, S, M, L, XL, and XXL. Each cap size covers an interval of head circumferences. The cap manufacturer gave students the table below that shows the interval of head circumferences for each cap size. The interval 510—< 530 represents head circumferences from 510 mm to 530 mm, not including 530.

Cap Sizes	Interval of Head Circumferences (millimeters)	Tally	Frequency
XS	510—< 530		
S	530—< 550		
M	550—< 570		
L	570—< 590		
XL	590—< 610		
XXL	610—< 630		

Exercises 1–4

1. What size cap would someone with a head circumference of 570 mm need?

2. Complete the tally and frequency columns in the table in Example 1 to determine the number of each size cap students need to order for the adults who wanted to order a cap.

3. What head circumference would you use to describe the center of the data?

4. Describe any patterns that you observe in the frequency column.

Example 2: Histogram

One student looked at the tally column and said that it looked somewhat like a bar graph turned on its side. A histogram is a graph that is like a bar graph except that the horizontal axis is a number line that is marked off in equal intervals.

To make a histogram:

- Draw a horizontal line, and mark the intervals.
- Draw a vertical line, and label it Frequency.
- Mark the Frequency axis with a scale that starts at 0 and goes up to something that is greater than the largest frequency in the frequency table.
- For each interval, draw a bar over that interval that has a height equal to the frequency for that interval.

The first two bars of the histogram have been drawn below.

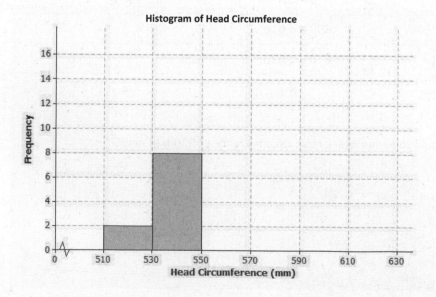

Histogram of Head Circumference

©2015 Great Minds eureka-math.org
G6-M5M6-SE-B3-1.3.1-01.2016

Exercises 5–9

5. Complete the histogram by drawing bars whose heights are the frequencies for the other intervals.

6. Based on the histogram, describe the center of the head circumferences.

7. How would the histogram change if you added head circumferences of 551 mm and 569 mm to the data set?

8. Because the 40 head circumference values were given, you could have constructed a dot plot to display the head circumference data. What information is lost when a histogram is used to represent a data distribution instead of a dot plot?

9. Suppose that there had been 200 head circumference measurements in the data set. Explain why you might prefer to summarize this data set using a histogram rather than a dot plot.

Example 3: Shape of the Histogram

A histogram is useful to describe the shape of the data distribution. It is important to think about the shape of a data distribution because depending on the shape, there are different ways to describe important features of the distribution, such as center and variability.

A group of students wanted to find out how long a certain brand of AA batteries lasted. The histogram below shows the data distribution for how long (in hours) that some AA batteries lasted. Looking at the shape of the histogram, notice how the data mound up around a center of approximately 105 hours. We would describe this shape as mound shaped or symmetric. If we were to draw a line down the center, notice how each side of the histogram is approximately the same, or a mirror image of the other. This means the histogram is approximately symmetrical.

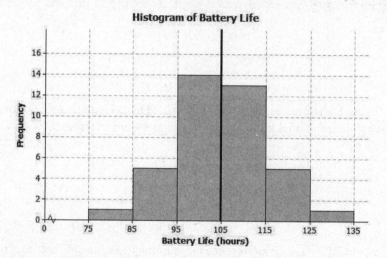

Another group of students wanted to investigate the maximum drop length for roller coasters. The histogram below shows the maximum drop (in feet) of a selected group of roller coasters. This histogram has a skewed shape. Most of the data are in the intervals from 50 feet to 170 feet. But there is one value that falls in the interval from 290 feet to 330 feet and one value that falls in the interval from 410 feet to 550 feet. These two values are unusual (or not typical) when compared to the rest of the data because they are much greater than most of the data.

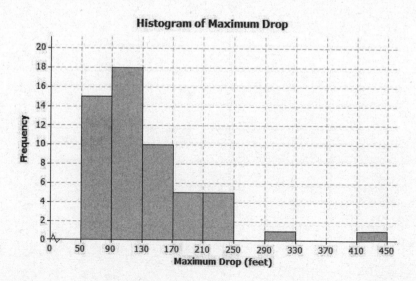

Exercises 10–12

10. The histogram below shows the highway miles per gallon of different compact cars.

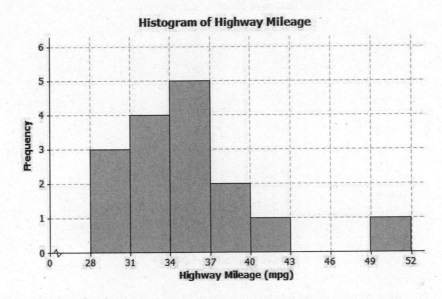

Histogram of Highway Mileage

a. Describe the shape of the histogram as approximately symmetric, skewed left, or skewed right.

b. Draw a vertical line on the histogram to show where the typical number of miles per gallon for a compact car would be.

c. What does the shape of the histogram tell you about miles per gallon for compact cars?

11. Describe the shape of the head circumference histogram that you completed in Exercise 5 as approximately symmetric, skewed left, or skewed right.

12. Another student decided to organize the head circumference data by changing the width of each interval to be 10 instead of 20. Below is the histogram that the student made.

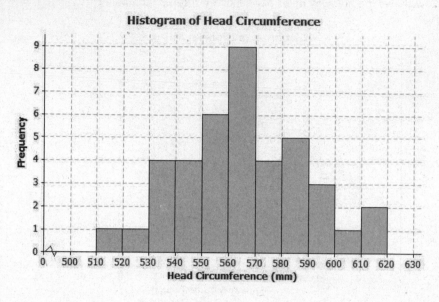

Histogram of Head Circumference

a. How does this histogram compare with the histogram of the head circumferences that you completed in Exercise 5?

b. Describe the shape of this new histogram as approximately symmetric, skewed left, or skewed right.

c. How many head circumferences are in the interval from 570 to 590 mm?

d. In what interval would a head circumference of 571 mm be included? In what interval would a head circumference of 610 mm be included?

EUREKA
MATH

Problem Set

1. The following histogram summarizes the ages of the actresses whose performances have won in the Best Leading Actress category at the annual Academy Awards (i.e., Oscars).

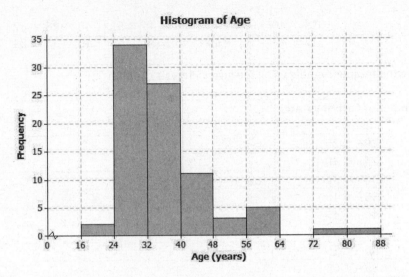

a. Which age interval contains the most actresses? How many actresses are represented in that interval?

b. Describe the shape of the histogram.

c. What does the histogram tell you about the ages of actresses who won the Oscar for best actress?

d. Which interval describes the center of the ages of the actresses?

e. An age of 72 would be included in which interval?

2. The frequency table below shows the seating capacity of arenas for NBA basketball teams.

Number of Seats	Tally	Frequency
17,000–< 17,500	||	2
17,500–< 18,000	|	1
18,000–< 18,500	ⅢⅢ |	6
18,500–< 19,000	ⅢⅢ	5
19,000–< 19,500	ⅢⅢ	5
19,500–< 20,000	ⅢⅢ	5
20,000–< 20,500	||	2
20,500–< 21,000	||	2
21,000–< 21,500		0
21,500–< 22,000		0
22,000–< 22,500	|	1

a. Draw a histogram for the number of seats in the NBA arenas data. Use the histograms you have seen throughout this lesson to help you in the construction of your histogram.

b. What is the width of each interval? How do you know?

c.   Describe the shape of the histogram.

d.   Which interval describes the center of the number of seats data?

3.   Listed are the grams of carbohydrates in hamburgers at selected fast food restaurants.

| 33 | 40 | 66 | 45 | 28 | 30 | 52 | 40 | 26 | 42 |
| 42 | 44 | 33 | 44 | 45 | 32 | 45 | 45 | 52 | 24 |

a.   Complete the frequency table using the given intervals of width 5.

Number of Carbohydrates (grams)	Tally	Frequency
20–< 25		
25–< 30		
30–< 35		
35–< 40		
40–< 45		
45–< 50		
50–< 55		
55–< 60		
60–< 65		
65–< 70		

b.   Draw a histogram of the carbohydrate data.

c.   Describe the center and shape of the histogram.

d.   In the frequency table below, the intervals are changed. Using the carbohydrate data above, complete the frequency table with intervals of width 10.

Number of Carbohydrates (grams)	Tally	Frequency
20–< 30		
30–< 40		
40–< 50		
50–< 60		
60–< 70		

e.   Draw a histogram.

4.   Use the histograms that you constructed in Exercise 3 parts (b) and (e) to answer the following questions.

a.   Why are there fewer bars in the histogram in part (e) than the histogram in part (b)?

b.   Did the shape of the histogram in part (e) change from the shape of the histogram in part (b)?

c.   Did your estimate of the center change from the histogram in part (b) to the histogram in part (e)?

EUREKA MATH

Lesson 5: Describing a Distribution Displayed in a Histogram

Classwork

Example 1: Relative Frequency Table

In Lesson 4, we investigated the head circumferences that the boys' and girls' basketball teams collected. Below is the frequency table of the head circumferences that they measured.

Cap Sizes	Interval of Head Circumferences (millimeters)	Tally	Frequency
XS	510−< 530	||	2
S	530−< 550	₩₩ |||	8
M	550−< 570	₩₩ ₩₩ ₩₩	15
L	570−< 590	₩₩ ||||	9
XL	590−< 610	||||	4
XXL	610−< 630	||	2
		Total:	40

Isabel, one of the basketball players, indicated that most of the caps were small (S), medium (M), or large (L). To decide if Isabel was correct, the players added a relative frequency column to the table.

Relative frequency is the frequency for an interval divided by the total number of data values. For example, the relative frequency for the extra small (XS) cap is 2 divided by 40, or 0.05. This represents the fraction of the data values that were XS.

Exercises 1–4

1. Complete the relative frequency column in the table below.

Cap Sizes	Interval of Head Circumferences (millimeters)	Tally	Frequency	Relative Frequency
XS	510–< 530	||	2	$\frac{2}{40} = 0.050$
S	530–< 550	⊬⊬ |||	8	$\frac{8}{40} = 0.200$
M	550–< 570	⊬⊬ ⊬⊬ ⊬⊬	15	
L	570–< 590	⊬⊬ ||||	9	
XL	590–< 610	||||	4	
XXL	610–< 630	||	2	
			Total: 40	

2. What is the total of the relative frequency column?

3. Which interval has the greatest relative frequency? What is the value?

4. What percentage of the head circumferences are between 530 and 589 mm? Show how you determined the answer.

EUREKA
MATH™

Example 2: Relative Frequency Histogram

The players decided to construct a histogram using the relative frequencies instead of the frequencies.

They noticed that the relative frequencies in the table ranged from close to 0 to about 0.40. They drew a number line and marked off the intervals on that line. Then, they drew the vertical line and labeled it Relative Frequency. They added a scale to this line by starting at 0 and counting by 0.05 until they reached 0.40.

They completed the histogram by drawing the bars so the height of each bar matched the relative frequency for that interval. Here is the completed relative frequency histogram:

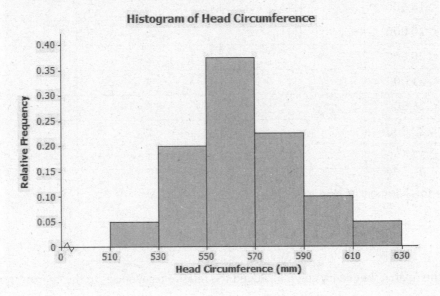

Exercises 5–6

5.

   a.   Describe the shape of the relative frequency histogram of head circumferences from Example 2.

   b.   How does the shape of this relative frequency histogram compare with the frequency histogram you drew in Exercise 5 of Lesson 4?

   c.   Isabel said that most of the caps that needed to be ordered were small (S), medium (M), and large (L). Was she right? What percentage of the caps to be ordered are small, medium, or large?

6. Here is the frequency table of the seating capacity of arenas for the NBA basketball teams.

Number of Seats	Tally	Frequency	Relative Frequency
17,000–< 17,500	||	2	
17,500–< 18,000	|	1	
18,000–< 18,500	₦₦₦ |	6	
18,500–< 19,000	₦₦₦	5	
19,000–< 19,500	₦₦₦	5	
19,500–< 20,000	₦₦₦	5	
20,000–< 20,500	||	2	
20,500–< 21,000	||	2	
21,000–< 21,500		0	
21,500–< 22,000		0	
22,000–< 22,500	|	1	

a. What is the total number of NBA arenas?

b. Complete the relative frequency column. Round the relative frequencies to the nearest thousandth.

c. Construct a relative frequency histogram.

d. Describe the shape of the relative frequency histogram.

e. What percentage of the arenas have a seating capacity between 18,500 and 19,999 seats?

f. How does this relative frequency histogram compare to the frequency histogram that you drew in Problem 2 of the Problem Set in Lesson 4?

Lesson Summary

A **relative frequency** is the frequency for an interval divided by the total number of data values. For example, if the first interval contains 8 out of a total of 32 data values, the relative frequency of the first interval is $\dfrac{8}{32} = \dfrac{1}{4} = 0.25$, or 25%.

A **relative frequency histogram** is a histogram that is constructed using relative frequencies instead of frequencies.

Problem Set

1. Below is a relative frequency histogram of the maximum drop (in feet) of a selected group of roller coasters.

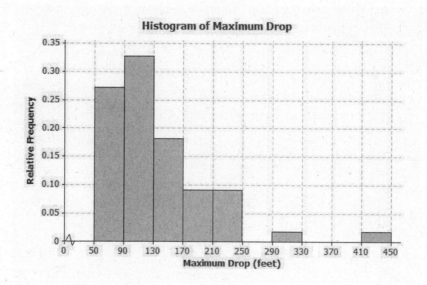

a.   Describe the shape of the relative frequency histogram.

b.   What does the shape tell you about the maximum drop (in feet) of roller coasters?

c.   Jerome said that more than half of the data values are in the interval from 50 to 130 feet. Do you agree with Jerome? Why or why not?

EUREKA MATH

2.   The frequency table below shows the length of selected movies shown in a local theater over the past 6 months.

Length of Movie (minutes)	Tally	Frequency	Relative Frequency
80–< 90	|	1	
90–< 100	||||	4	
100–< 110	⦀ ||	7	
110–< 120	⦀	5	
120–< 130	⦀ ||	7	
130–< 140	|||	3	
140–< 150	|	1	

a.   Complete the relative frequency column. Round the relative frequencies to the nearest thousandth.

b.   What percentage of the movie lengths are greater than or equal to 130 minutes?

c.   Draw a relative frequency histogram. (Hint: Label the relative frequency scale starting at 0 and going up to 0.30, marking off intervals of 0.05.)

d.   Describe the shape of the relative frequency histogram.

e.   What does the shape tell you about the length of movie times?

3.   The table below shows the highway miles per gallon of different compact cars.

Mileage	Tally	Frequency	Relative Frequency
28–< 31	|||	3	
31–< 34	||||	4	
34–< 37	⦀	5	
37–< 40	||	2	
40–< 43	|	1	
43–< 46		0	
46–< 49		0	
49–< 52	|	1	

a.   What is the total number of compact cars?

b.   Complete the relative frequency column. Round the relative frequencies to the nearest thousandth.

c.   What percent of the cars get between 31 and up to but not including 37 miles per gallon on the highway?

d.   Juan drew the relative frequency histogram of the highway miles per gallon for the compact cars, shown on the right. Did Juan draw the histogram correctly? Explain your answer.

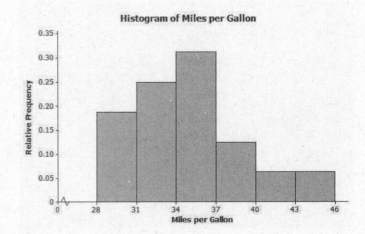

This page intentionally left blank

Lesson 6: Describing the Center of a Distribution Using the Mean

Classwork

Example 1

Recall that in Lesson 3, Robert, a sixth grader at Roosevelt Middle School, investigated the number of hours of sleep sixth-grade students get on school nights. Today, he is to make a short report to the class on his investigation. Here is his report.

"I took a survey of twenty-nine sixth graders, asking them, 'How many hours of sleep per night do you usually get when you have school the next day?' The first thing I had to do was to organize the data. I did this by drawing a dot plot. Looking at the dot plot, I would say that a typical amount of sleep is 8 or 9 hours."

Dot Plot of Number of Hours of Sleep

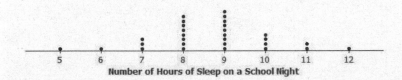

Number of Hours of Sleep on a School Night

Michelle is Robert's classmate. She liked his report but has a really different thought about determining the center of the number of hours of sleep. Her idea is to even out the data in order to determine a typical or center value.

Exercises 1–6

Suppose that Michelle asks ten of her classmates for the number of hours they usually sleep when there is school the next day.

Suppose they responded (in hours): 8 10 8 8 11 11 9 8 10 7.

1. How do you think Robert would organize this new data? What do you think Robert would say is the center of these ten data points? Why?

2. Do you think his value is a good measure to use for the center of Michelle's data set? Why or why not?

The measure of center that Michelle is proposing is called the *mean*. She finds the total number of hours of sleep for the ten students. That is 90 hours. She has 90 Unifix cubes (Snap cubes). She gives each of the ten students the number of cubes that equals the number of hours of sleep each had reported. She then asks each of the ten students to connect their cubes in a stack and put their stacks on a table to compare them. She then has them share their cubes with each other until they all have the same number of cubes in their stacks when they are done sharing.

3. Make ten stacks of cubes representing the number of hours of sleep for each of the ten students. Using Michelle's method, how many cubes are in each of the ten stacks when they are done sharing?

4. Noting that each cube represents one hour of sleep, interpret your answer to Exercise 3 in terms of number of hours of sleep. What does this number of cubes in each stack represent? What is this value called?

5. Suppose that the student who told Michelle he slept 7 hours changes his data value to 8 hours. What does Michelle's procedure now produce for her center of the new set of data? What did you have to do with that extra cube to make Michelle's procedure work?

6. Interpret Michelle's fair share procedure by developing a mathematical formula that results in finding the fair share value without actually using cubes. Be sure that you can explain clearly how the fair share procedure and the mathematical formula relate to each other.

©2015 Great Minds eureka-math.org
G6-M5M6-SE-B3-1.3.1-01.2016

Suppose that Robert asked five sixth graders how many pets each had. Their responses were 2, 6, 2, 4, 1. Robert showed the data with cubes as follows:

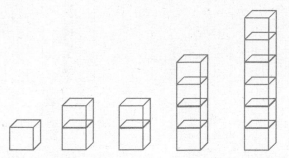

Note that one student has one pet, two students have two pets each, one student has four pets, and one student has six pets. Robert also represented the data set in the following dot plot.

Dot Plot of Number of Pets

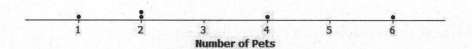

Number of Pets

Robert wants to illustrate Michelle's fair share method by using dot plots. He drew the following dot plot and said that it represents the result of the student with six pets sharing one of her pets with the student who has one pet.

Dot Plot of Number of Pets

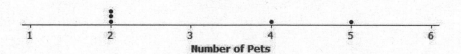

Number of Pets

©2015 Great Minds eureka-math.org
G6-M5M6-SE-B3-1.3.1-01.2016

Robert also represented the dot plot above with cubes. His representation is shown below.

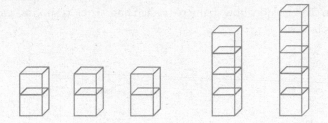

Exercises 7–10

Now, continue distributing the pets based on the following steps.

7. Robert does a fair share step by having the student with five pets share one of her pets with one of the students with two pets.

   a. Draw the cubes representation that shows the result of this fair share step.

   b. Draw the dot plot that shows the result of this fair share step.

8. Robert does another fair share step by having one of the students who has four pets share one pet with one of the students who has two pets.

   a. Draw the cubes representation that shows the result of this fair share step.

Lesson 6:   Describing the Center of a Distribution Using the Mean

Problem Set

1. The number of pockets in the clothes worn by four students to school today is 4, 1, 3, 4.

   a. Perform the fair share process to find the mean number of pockets for these four students. Sketch the cubes representations for each step of the process.

   b. Find the total of the distances on each side of the mean to show the mean found in part (a) is correct.

2. The times (rounded to the nearest minute) it took each of six classmates to run a mile are 7, 9, 10, 11, 11, and 12 minutes.

   a. Draw a dot plot representation for the mile times.

   b. Suppose that Sabina thinks the mean is 11 minutes. Is she correct? Explain your answer.

   c. What is the mean?

3. The prices per gallon of gasoline (in cents) at five stations across town on one day are shown in the following dot plot. The price for a sixth station is missing, but the mean price for all six stations was reported to be 380 cents per gallon. Use the balancing process to determine the price of a gallon of gasoline at the sixth station.

   Dot Plot of Price (cents per gallon)

   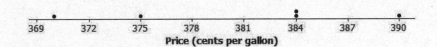

   Price (cents per gallon)

4. The number of phones (landline and cell) owned by the members of each of nine families is 3, 5, 6, 6, 6, 6, 7, 7, 8.

   a. Use the mathematical formula for the mean (determine the sum of the data points, and divide by the number of data points) to find the mean number of phones owned for these nine families.

   b. Draw a dot plot of the data, and verify your answer in part (a) by using the balancing process.

This page intentionally left blank

Lesson 8: Variability in a Data Distribution

Classwork

Example 1: Comparing Two Data Distributions

Robert's family is planning to move to either New York City or San Francisco. Robert has a cousin in San Francisco and asked her how she likes living in a climate as warm as San Francisco. She replied that it doesn't get very warm in San Francisco. He was surprised by her answer. Because temperature was one of the criteria he was going to use to form his opinion about where to move, he decided to investigate the temperature distributions for New York City and San Francisco. The table below gives average temperatures (in degrees Fahrenheit) for each month for the two cities.

City	Jan.	Feb.	Mar.	Apr.	May	June	July	Aug.	Sep.	Oct.	Nov.	Dec.
New York City	39	42	50	61	71	81	85	84	76	65	55	47
San Francisco	57	60	62	63	64	67	67	68	70	69	63	58

Data Source as of 2013: http://www.usclimatedata.com/climate/san-francisco/california/united-states/usca0987

Data Source as of 2013: http://www.usclimatedata.com/climate/new-york/united-states/3202

Exercises 1–2

Use the data in the table provided in Example 1 to answer the following:

1. Calculate the mean of the monthly average temperatures for each city.

2. Recall that Robert is trying to decide where he wants to move. What is your advice to him based on comparing the means of the monthly temperatures of the two cities?

Example 2: Understanding Variability

Maybe Robert should look at how spread out the New York City monthly temperature data are from the mean of the New York City monthly temperatures and how spread out the San Francisco monthly temperature data are from the mean of the San Francisco monthly temperatures. To compare the variability of monthly temperatures between the two cities, it may be helpful to look at dot plots. The dot plots of the monthly temperature distributions for New York City and San Francisco follow.

Dot Plot of Temperature for New York City   Dot Plot of Temperature for San Francisco

Exercises 3–7

Use the dot plots above to answer the following:

3. Mark the location of the mean on each distribution with the balancing Δ symbol. How do the two distributions compare based on their means?

4. Describe the variability of the New York City monthly temperatures from the New York City mean.

5. Describe the variability of the San Francisco monthly temperatures from the San Francisco mean.

EUREKA
MATH

6. Compare the variability in the two distributions. Is the variability about the same, or is it different? If different, which monthly temperature distribution has more variability? Explain.

7. If Robert prefers to choose the city where the temperatures vary the least from month to month, which city should he choose? Explain.

Example 3: Considering the Mean and Variability in a Data Distribution

The mean is used to describe a typical value for the entire data distribution. Sabina asks Robert which city he thinks has the better climate. How do you think Robert responds?

Sabina is confused and asks him to explain what he means by this statement. How could Robert explain what he means?

Exercises 8–14

Consider the following two distributions of times it takes six students to get to school in the morning and to go home from school in the afternoon.

	Time (minutes)					
Morning	11	12	14	14	16	17
Afternoon	6	10	13	18	18	19

8. To visualize the means and variability, draw a dot plot for each of the two distributions.

Morning

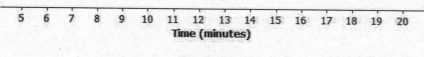

Afternoon

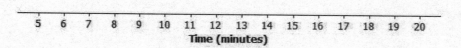

9. What is the mean time to get from home to school in the morning for these six students?

10. What is the mean time to get from school to home in the afternoon for these six students?

11. For which distribution does the mean give a more accurate indicator of a typical time? Explain your answer.

Distributions can be ordered according to how much the data values vary around their means.

Consider the following data on the number of green jelly beans in seven bags of jelly beans from each of five different candy manufacturers (AllGood, Best, Delight, Sweet, and Yum). The mean in each distribution is 42 green jelly beans.

	Bag 1	Bag 2	Bag 3	Bag 4	Bag 5	Bag 6	Bag 7
AllGood	40	40	41	42	42	43	46
Best	22	31	36	42	48	53	62
Delight	26	36	40	43	47	50	52
Sweet	36	39	42	42	42	44	49
Yum	33	36	42	42	45	48	48

12. Draw a dot plot of the distribution of the number of green jelly beans for each of the five candy makers. Mark the location of the mean on each distribution with the balancing Δ symbol.

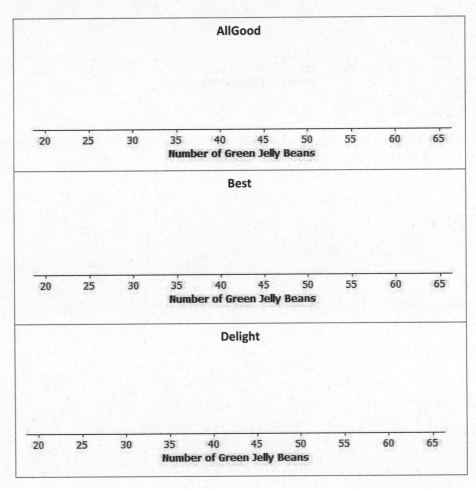

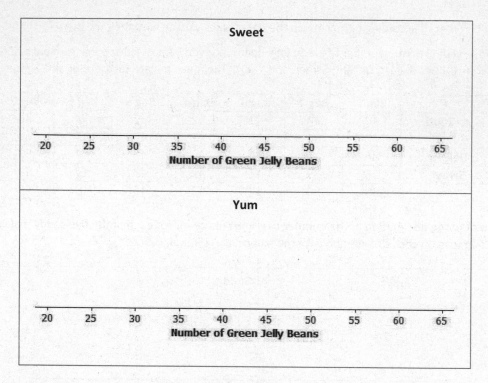

13. Order the candy manufacturers from the one you think has the least variability to the one with the most variability. Explain your reasoning for choosing the order.

14. For which company would the mean be considered a better indicator of a typical value (based on least variability)?

©2015 Great Minds eureka-math.org
G6-M5M6-SE-B3-1.3.1-01.2016

Lesson Summary

We can compare distributions based on their means, but variability must also be considered. The mean of a distribution with small variability (not a lot of spread) is considered to be a better indication of a typical value than the mean of a distribution with greater variability (or wide spread).

Problem Set

1. The number of pockets in the clothes worn by seven students to school yesterday was 4, 1, 3, 4, 2, 2, 5. Today, those seven students each had three pockets in their clothes.

   a. Draw one dot plot of the number of pockets data for what students wore yesterday and another dot plot for what students wore today. Be sure to use the same scale.

   b. For each distribution, find the mean number of pockets worn by the seven students. Show the means on the dot plots by using the balancing Δ symbol.

   c. For which distribution is the mean number of pockets a better indicator of what is typical? Explain.

2. The number of minutes (rounded to the nearest minute) it took to run a certain route was recorded for each of five students. The resulting data were 9, 10, 11, 14, and 16 minutes. The number of minutes (rounded to the nearest minute) it took the five students to run a different route was also recorded, resulting in the following data: 6, 8, 12, 15, and 19 minutes.

   a. Draw dot plots for the distributions of the times for the two routes. Be sure to use the same scale on both dot plots.

   b. Do the distributions have the same mean? What is the mean of each dot plot?

   c. In which distribution is the mean a better indicator of the typical amount of time taken to run the route? Explain.

3. The following table shows the prices per gallon of gasoline (in cents) at five stations across town as recorded on Monday, Wednesday, and Friday of a certain week.

Day	R&C	Al's	PB	Sam's	Ann's
Monday	359	358	362	359	362
Wednesday	357	365	364	354	360
Friday	350	350	360	370	370

   a. The mean price per day for the five stations is the same for each of the three days. Without doing any calculations and simply looking at Friday's prices, what must the mean price be?

   b. For which daily distribution is the mean a better indicator of the typical price per gallon for the five stations? Explain.

Lesson 8:   Variability in a Data Distribution   S.57

©2015 Great Minds eureka-math.org
G6-M5M6-SE-B3-1.3.1-01.2016

This page intentionally left blank

Lesson 9: The Mean Absolute Deviation (MAD)

Classwork

Example 1: Variability

In Lesson 8, Robert wanted to decide where he would rather move (New York City or San Francisco). He planned to make his decision by comparing the average monthly temperatures for the two cities. Since the mean of the average monthly temperatures for New York City and the mean for San Francisco turned out to be about the same, he decided instead to compare the cities based on the variability in their monthly average temperatures. He looked at the two distributions and decided that the New York City temperatures were more spread out from their mean than were the San Francisco temperatures from their mean.

Exercises 1–3

The following temperature distributions for seven other cities all have a mean monthly temperature of approximately 63 degrees Fahrenheit. They do not have the same variability.

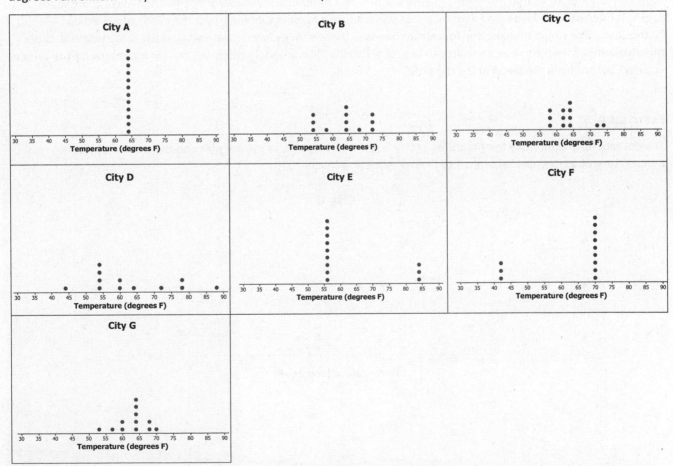

©2015 Great Minds eureka-math.org
G6-M5M6-SE-B3-1.3.1-01.2016

1. Which distribution has the smallest variability? Explain your answer.

2. Which distribution or distributions seem to have the most variability? Explain your answer.

3. Order the seven distributions from least variability to most variability. Explain why you listed the distributions in the order that you chose.

Example 2: Measuring Variability

Based on just looking at the distributions, there are different orderings of variability that seem to make some sense. Sabina is interested in developing a formula that will produce a number that measures the variability in a data distribution. She would then use the formula to measure the variability in each data set and use these values to order the distributions from smallest variability to largest variability. She proposes beginning by looking at how far the values in a data set are from the mean of the data set.

Exercises 4–5

The dot plot for the monthly temperatures in City G is shown below. Use the dot plot and the mean monthly temperature of 63 degrees Fahrenheit to answer the following questions.

City G

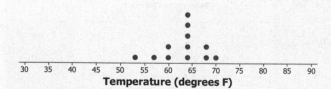

Temperature (degrees F)

4. Fill in the following table for City G's temperature deviations.

Temperature (in degrees Fahrenheit)	Distance (in degrees Fahrenheit) from the Mean of 63°F	Deviation from the Mean (distance and direction)
53	10	10 to the left
57		
60		
60		
64		
64		
64		
64		
64		
68		
68		
70		

5. What is the sum of the distances to the left of the mean? What is the sum of the distances to the right of the mean?

©2015 Great Minds eureka-math.org
G6-M5M6-SE-B3-1.3.1-01.2016

Sabina notices that when there is not much variability in a data set, the distances from the mean are small and that when there is a lot of variability in a data set, the data values are spread out and at least some of the distances from the mean are large. She wonders how she can use the distances from the mean to help her develop a formula to measure variability.

Exercises 6–7

6. Use the data on monthly temperatures for City G given in Exercise 4 to answer the following questions.

   a.   Fill in the following table.

Temperature (in degrees Fahrenheit)	Distance from the Mean (absolute deviation)
53	10
57	
60	
60	
64	
64	
64	
64	
64	
68	
68	
70	

   b.   The absolute deviation for a data value is its distance from the mean of the data set. For example, for the first temperature value for City G (53 degrees), the absolute deviation is 10. What is the sum of the absolute deviations?

c. Sabina suggests that the mean of the absolute deviations (the mean of the distances) could be a measure of the variability in a data set. Its value is the average distance of the data values from the mean of the monthly temperatures. It is called the *mean absolute deviation* and is denoted by the letters MAD. Find the MAD for this data set of City G's temperatures. Round to the nearest tenth.

d. Find the MAD values in degrees Fahrenheit for each of the seven city temperature distributions, and use the values to order the distributions from least variability to most variability. Recall that the mean for each data set is 63 degrees Fahrenheit. Looking only at the distributions, does the list that you made in Exercise 2 match the list made by ordering MAD values?

> **MAD values (in °F):**

e. Which of the following is a correct interpretation of the MAD?

    i.   The monthly temperatures in City G are all within 3.7 degrees from the approximate mean of 63 degrees.

    ii. The monthly temperatures in City G are, on average, 3.7 degrees from the approximate mean temperature of 63 degrees.

    iii. All of the monthly temperatures in City G differ from the approximate mean temperature of 63 degrees by 3.7 degrees.

7.   The dot plot for City A's temperatures follows.

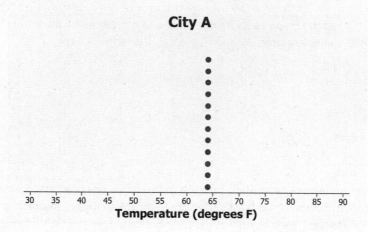

a.   How much variability is there in City A's temperatures? Why?

b.   Does the MAD agree with your answer in part (a)?

EUREKA
MATH™

Lesson Summary

In this lesson, a formula was developed that measures the amount of variability in a data distribution.

- The absolute deviation of a data point is the distance that data point is from the mean.
- The mean absolute deviation (MAD) is computed by finding the mean of the absolute deviations (distances from the mean) for the data set.
- The value of MAD is the average distance that the data values are from the mean.
- A small MAD indicates that the data distribution has very little variability.
- A large MAD indicates that the data points are spread out and that at least some are far away from the mean.

Problem Set

1. Suppose the dot plot on the left shows the number of goals a boys' soccer team has scored in six games so far this season, and the dot plot on the right shows the number of goals a girls' soccer team has scored in six games so far this season. The mean for both of these teams is 3.

Dot Plot of Number of Goals Scored for Boys' Team   Dotplot of Number of Goals Scored for Girls' Team

Number of Goals Scored   Number of Goals Scored

a. Before doing any calculations, which dot plot has the larger MAD? Explain how you know.

b. Use the following tables to find the MAD for each distribution. Round your calculations to the nearest hundredth.

Boys' Team	
Number of Goals	Absolute Deviation
0	
0	
3	
3	
5	
7	
Sum	

Girls' Team	
Number of Goals	Absolute Deviation
2	
2	
3	
3	
3	
5	
Sum	

c. Based on the computed MAD values, for which distribution is the mean a better indication of a typical value? Explain your answer.

©2015 Great Minds eureka-math.org
G6-M5M6-SE-B3-1.3.1-01.2016

2. Recall Robert's problem of deciding whether to move to New York City or to San Francisco. A table of temperatures (in degrees Fahrenheit) and absolute deviations for New York City follows:

Average Temperature in New York City												
Month	Jan.	Feb.	Mar.	Apr.	May	June	July	Aug.	Sep.	Oct.	Nov.	Dec.
Temperature (°F)	39	42	50	61	71	81	85	84	76	65	55	47
Absolute Deviation	24	21	13	2	8	18	22	21	13	2	8	16

a. The absolute deviations for the monthly temperatures are shown in the above table. Use this information to calculate the MAD. Explain what the MAD means in words.

b. Complete the following table, and then use the values to calculate the MAD for the San Francisco data distribution.

Average Temperature in San Francisco												
Month	Jan.	Feb.	Mar.	Apr.	May	June	July	Aug.	Sep.	Oct.	Nov.	Dec.
Temperature (°F)	57	60	62	63	64	67	67	68	70	69	63	58
Absolute Deviation												

c. Comparing the MAD values for New York City and San Francisco, which city would Robert choose to move to if he is interested in having a lot of variability in monthly temperatures? Explain using the MAD.

3. Consider the following data of the number of green jelly beans in seven bags sampled from each of five different candy manufacturers (Awesome, Delight, Finest, Sweeties, YumYum). Note that the mean of each distribution is 42 green jelly beans.

	Bag 1	Bag 2	Bag 3	Bag 4	Bag 5	Bag 6	Bag 7
Awesome	40	40	41	42	42	43	46
Delight	22	31	36	42	48	53	62
Finest	26	36	40	43	47	50	52
Sweeties	36	39	42	42	42	44	49
YumYum	33	36	42	42	45	48	48

a. Complete the following table of the absolute deviations for the seven bags for each candy manufacturer.

Absolute Deviations							
	Bag 1	Bag 2	Bag 3	Bag 4	Bag 5	Bag 6	Bag 7
Awesome	2	2	1	0	0	1	4
Delight	20	11	6				
Finest	16						
Sweeties							
YumYum							

EUREKA
MATH™

b. Based on what you learned about MAD, which manufacturer do you think will have the lowest MAD? Calculate the MAD for the manufacturer you selected.

	Bag 1	Bag 2	Bag 3	Bag 4	Bag 5	Bag 6	Bag 7	SUM	MAD
Awesome									
Delight									
Finest									
Sweeties									
YumYum									

©2015 Great Minds eureka-math.org
G6-M5M6-SE-B3-1.3.1-01.2016

This page intentionally left blank

Lesson 10: Describing Distributions Using the Mean and MAD

Classwork

Example 1: Describing Distributions

In Lesson 9, Sabina developed the mean absolute deviation (MAD) as a number that measures variability in a data distribution. Using the mean and MAD along with a dot plot allows you to describe the center, spread, and shape of a data distribution. For example, suppose that data on the number of pets for ten students are shown in the dot plot below.

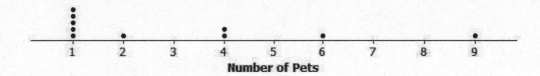

There are several ways to describe the data distribution. The mean number of pets for these students is 3, which is a measure of center. There is variability in the number of pets the students have, and data values differ from the mean by about 2.2 pets on average (the MAD). The shape of the distribution is heavy on the left, and then it thins out to the right.

Exercises 1–4

1. Suppose that the weights of seven middle school students' backpacks are given below.

   a. Fill in the following table.

Student	Alan	Beth	Char	Damon	Elisha	Fred	Georgia
Weight (pounds)	18	18	18	18	18	18	18
Deviation							
Absolute Deviation							

   b. Draw a dot plot for these data, and calculate the mean and MAD.

c.   Describe this distribution of weights of backpacks by discussing the center, spread, and shape.

2.   Suppose that the weight of Elisha's backpack is 17 pounds rather than 18 pounds.

a.   Draw a dot plot for the new distribution.

b.   Without doing any calculations, how is the mean affected by the lighter weight? Would the new mean be the same, smaller, or larger?

c.   Without doing any calculations, how is the MAD affected by the lighter weight? Would the new MAD be the same, smaller, or larger?

3.   Suppose that in addition to Elisha's backpack weight having changed from 18 to 17 pounds, Fred's backpack weight is changed from 18 to 19 pounds.

a.   Draw a dot plot for the new distribution.

©2015 Great Minds eureka-math.org
G6-M5M6-SE-B3-1.3.1-01.2016

EUREKA
MATH™

b. Without doing any calculations, how would the new mean compare to the original mean?

c. Without doing any calculations, would the MAD for the new distribution be the same as, smaller than, or larger than the original MAD?

d. Without doing any calculations, how would the MAD for the new distribution compare to the one in Exercise 2?

4. Suppose that seven second graders' backpack weights were as follows:

Student	Alice	Bob	Carol	Damon	Ed	Felipe	Gale
Weight (pounds)	5	5	5	5	5	5	5

a. How is the distribution of backpack weights for the second graders similar to the original distribution for the middle school students given in Exercise 1?

b. How are the distributions different?

©2015 Great Minds eureka-math.org
G6-M5M6-SE-B3-1.3.1-01.2016

Example 2: Using the MAD

Using data to make decisions often involves comparing distributions. Recall that Robert is trying to decide whether to move to New York City or to San Francisco based on temperature. Comparing the center, spread, and shape for the two temperature distributions could help him decide.

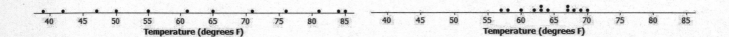

Dot Plot of Temperature for New York City   Dot Plot of Temperature for San Francisco

From the dot plots, Robert saw that monthly temperatures in New York City were spread fairly evenly from around 40 degrees to around 85 degrees, but in San Francisco, the monthly temperatures did not vary as much. He was surprised that the mean temperature was about the same for both cities. The MAD of 14 degrees for New York City told him that, on average, a month's temperature was 14 degrees away from the mean of 63 degrees. That is a lot of variability, which is consistent with the dot plot. On the other hand, the MAD for San Francisco told him that San Francisco's monthly temperatures differ, on average, only 3.5 degrees from the mean of 64 degrees. So, the mean doesn't help Robert very much in making a decision, but the MAD and dot plot are helpful.

Which city should he choose if he loves warm weather and really dislikes cold weather?

Exercises 5–7

5. Robert wants to compare temperatures in degrees Fahrenheit for Cities B and C.

	Jan.	Feb.	Mar.	Apr.	May	June	July	Aug.	Sept.	Oct.	Nov.	Dec.
City B	54	54	58	63	63	68	72	72	72	63	63	54
City C	54	44	54	61	63	72	78	85	78	59	54	54

   a. Draw a dot plot of the monthly temperatures for each of the cities.

b. Verify that the mean monthly temperature for each distribution is 63 degrees.

c. Find the MAD for each of the cities. Interpret the two MADs in words, and compare their values. Round your answers to the nearest tenth of a degree.

6. How would you describe the differences in the shapes of the monthly temperature distributions of the two cities?

7. Suppose that Robert had to decide between Cities D, E, and F.

	Jan.	Feb.	Mar.	Apr.	May	June	July	Aug.	Sept.	Oct.	Nov.	Dec.	Mean	MAD
City D	54	44	54	59	63	72	78	87	78	59	54	54	63	10.5
City E	56	56	56	56	56	84	84	84	56	56	56	56	63	10.5
City F	42	42	70	70	70	70	70	70	70	70	70	42	63	10.5

a. Draw a dot plot for each distribution.

b. Interpret the MAD for the distributions. What does this mean about variability?

c. How will Robert decide to which city he should move? List possible reasons Robert might have for choosing each city.

©2015 Great Minds eureka-math.org
G6-M5M6-SE-B3-1.3.1-01.2016

Lesson Summary

A data distribution can be described in terms of its center, spread, and shape.

- The center can be measured by the mean.
- The spread can be measured by the mean absolute deviation (MAD).
- A dot plot shows the shape of the distribution.

Problem Set

1. Draw a dot plot of the times that five students studied for a test if the mean time they studied was 2 hours and the MAD was 0 hours.

2. Suppose the times that five students studied for a test are as follows:

Student	Aria	Ben	Chloe	Dellan	Emma
Time (hours)	1.5	2	2	2.5	2

Michelle said that the MAD for this data set is 0 hours because the dot plot is balanced around 2. Without doing any calculations, do you agree with Michelle? Why or why not?

3. Suppose that the number of text messages eight students receive on a typical day is as follows:

Student	1	2	3	4	5	6	7	8
Number of Text Messages	42	56	35	70	56	50	65	50

a. Draw a dot plot for the number of text messages received on a typical day for these eight students.

b. Find the mean number of text messages these eight students receive on a typical day.

c. Find the MAD for the number of text messages, and explain its meaning using the words of this problem.

d. Describe the shape of this data distribution.

e. Suppose that in the original data set, Student 3 receives an additional five text messages per day, and Student 4 receives five fewer text messages per day.

   i. Without doing any calculations, does the mean for the new data set stay the same, increase, or decrease as compared to the original mean? Explain your reasoning.

   ii. Without doing any calculations, does the MAD for the new data set stay the same, increase, or decrease as compared to the original MAD? Explain your reasoning.

©2015 Great Minds eureka-math.org
G6-M5M6-SE-B3-1.3.1-01.2016

This page intentionally left blank

Lesson 11: Describing Distributions Using the Mean and MAD

Classwork

Example 1: Comparing Distributions with the Same Mean

In Lesson 10, a data distribution was characterized mainly by its center (mean) and variability (MAD). How these measures help us make a decision often depends on the context of the situation. For example, suppose that two classes of students took the same test, and their grades (based on 100 points) are shown in the following dot plots. The mean score for each distribution is 79 points. Would you rather be in Class A or Class B if you had a score of 79?

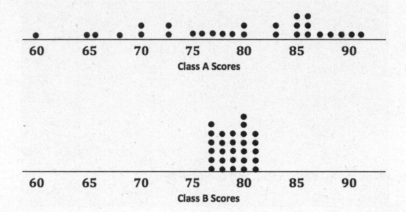

Exercises 1–6

1. Looking at the dot plots, which class has the greater MAD? Explain without actually calculating the MAD.

2. If Liz had one of the highest scores in her class, in which class would she rather be? Explain your reasoning.

3. If Logan scored below average, in which class would he rather be? Explain your reasoning.

©2015 Great Minds eureka-math.org
G6-M5M6-SE-B3-1.3.1-01.2016

Your little brother asks you to replace the battery in his favorite remote control car. The car is constructed so that it is difficult to replace its battery. Your research of the lifetimes (in hours) of two different battery brands (A and B) shows the following lifetimes for 20 batteries from each brand:

A	12	14	14	15	16	17	17	18	19	20	21	21	23	23	24	24	24	25	26	27
B	18	18	19	19	19	19	19	19	20	20	20	20	20	21	21	21	21	22	22	22

4. To help you decide which battery to purchase, start by drawing a dot plot of the lifetimes for each brand.

5. Find the mean battery lifetime for each brand, and compare them.

6. Looking at the variability in the dot plot for each data set, give one reason you might choose Brand A. What is one reason you might choose Brand B? Explain your reasoning.

Example 2: Comparing Distributions with Different Means

You have been comparing distributions that have the same mean but different variability. As you have seen, deciding whether large variability or small variability is best depends on the context and on what is being asked. If two data distributions have different means, do you think that variability will still play a part in making decisions?

EUREKA
MATH

©2015 Great Minds eureka-math.org
G6-M5M6-SE-B3-1.3.1-01.2016

Exercises 7–9

Suppose that you wanted to answer the following question: Are field crickets better predictors of air temperature than katydids? Both species of insect make chirping sounds by rubbing their front wings together.

The following data are the number of chirps (per minute) for 10 insects of each type. All the data were taken on the same evening at the same time.

Insect	1	2	3	4	5	6	7	8	9	10
Crickets	35	32	35	37	34	34	38	35	36	34
Katydids	66	62	61	64	63	62	68	64	66	64

7. Draw dot plots for these two data distributions using the same scale, going from 30 to 70. Visually, what conclusions can you draw from the dot plots?

8. Calculate the mean and MAD for each distribution.

9. The outside temperature T, in degrees Fahrenheit, can be predicted by using two different formulas. The formulas include the mean number of chirps per minute made by crickets or katydids.

    a.   For crickets, T is predicted by adding 40 to the mean number of chirps per minute. What value of T is being predicted by the crickets?

    b.   For katydids, T is predicted by adding 161 to the mean number of chirps per minute and then dividing the sum by 3. What value of T is being predicted by the katydids?

    c.   The temperature was 75 degrees Fahrenheit when these data were recorded, so using the mean from each data set gave an accurate prediction of temperature. If you were going to use the number of chirps from a single cricket or a single katydid to predict the temperature, would you use a cricket or a katydid? Explain how variability in the distributions of number of chirps played a role in your decision.

EUREKA
MATH™

Problem Set

1. Two classes took the same mathematics test. Summary measures for the two classes are as follows:

	Mean	MAD
Class A	78	2
Class B	78	10

  a. Suppose that you received the highest score in your class. Would your score have been higher if you were in Class A or Class B? Explain your reasoning.

  b. Suppose that your score was below the mean score. In which class would you prefer to have been? Explain your reasoning.

2. Eight of each of two varieties of tomato plants, LoveEm and Wonderful, are grown under the same conditions. The numbers of tomatoes produced from each plant of each variety are shown:

Plant	1	2	3	4	5	6	7	8
LoveEm	27	29	27	28	31	27	28	27
Wonderful	31	20	25	50	32	25	22	51

  a. Draw dot plots to help you decide which variety is more productive.

  b. Calculate the mean number of tomatoes produced for each variety. Which one produces more tomatoes on average?

  c. If you want to be able to accurately predict the number of tomatoes a plant is going to produce, which variety should you choose—the one with the smaller MAD or the one with the larger MAD? Explain your reasoning.

  d. Calculate the MAD of each plant variety.

This page intentionally left blank

Lesson 12: Describing the Center of a Distribution Using the Median

Classwork

How do we summarize a data distribution? What provides us with a good description of the data? The following exercises help us to understand how a numerical summary provides an answer to these questions.

Example 1: The Median—A Typical Number

Suppose a chain restaurant (Restaurant A) advertises that a typical number of french fries in a large bag is 82. The dot plot shows the number of french fries in a sample of twenty large bags from Restaurant A.

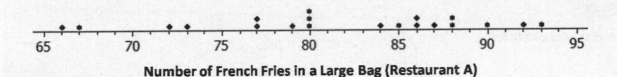

Number of French Fries in a Large Bag (Restaurant A)

Sometimes it is useful to know what point separates a data distribution into two equal parts, where one part represents the upper half of the data values and the other part represents the lower half of the data values. This point is called the *median*. When the data are arranged in order from smallest to largest, the same number of values will be above the median point as below the median.

Exercises 1–3

1. You just bought a large bag of fries from the restaurant. Do you think you have exactly 82 french fries? Why or why not?

2. How many bags were in the sample?

3. Which of the following statement(s) would seem to be true for the given data? Explain your reasoning.

    a. Half of the bags had more than 82 fries in them.

    b. Half of the bags had fewer than 82 fries in them.

    c. More than half of the bags had more than 82 fries in them.

    d. More than half of the bags had fewer than 82 fries in them.

    e. If you got a random bag of fries, you could get as many as 93 fries.

Example 2

Examine the dot plot below.

Grades on a Science Test

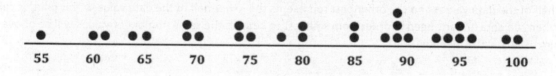

55   60   65   70   75   80   85   90   95   100

    a. How many data values are represented on the dot plot above?

    b. How many data values should be located above the median? How many below the median? Explain.

    c. For this data set, 14 values are 80 or smaller, and 14 values are 85 or larger, so the median should be between 80 and 85. When the median falls between two values in a data set, we use the average of the two middle values. For this example, the two middle values are 80 and 85. What is the median of the data presented on the dot plot?

©2015 Great Minds eureka-math.org
G6-M5M6-SE-B3-1.3.1-01.2016

d.   What does this information tell us about the data?

Example 3

Use the information from the dot plot in Example 2.

a.   What percentage of students scored higher than the median? Lower than the median?

b.   Suppose the teacher made a mistake, and the student who scored 65 actually scored a 71. Would the median change? Why or why not?

c.   Suppose the student who scored a 65 actually scored an 89. Would the median change? Why or why not?

Example 4

A grocery store usually has three checkout lines open on Saturday afternoons. One Saturday afternoon, the store manager decided to count how many customers were waiting to check out at 10 different times. She calculated the median of her ten data values to be 8 customers.

a.   Why might the median be an important number for the store manager to consider?

b. Give another example of when the median of a data set might provide useful information. Explain your thinking.

Exercises 4–5: A Skewed Distribution

4. The owner of the chain decided to check the number of french fries at another restaurant in the chain. Here are the data for Restaurant B: 82, 83, 83, 79, 85, 82, 78, 76, 76, 75, 78, 74, 70, 60, 82, 82, 83, 83, 83.

a. How many bags of fries were counted?

b. Sallee claims the median is 75 because she sees that 75 is the middle number in the data set listed above. She thinks half of the bags had fewer than 75 fries because there are 9 data values that come before 75 in the list, and there are 9 data values that come after 75 in the list. Do you think she would change her mind if the data were plotted in a dot plot? Why or why not?

c. Jake said the median was 83. What would you say to Jake?

d. Betse argued that the median was halfway between 60 and 85, or 72.5. Do you think she is right? Why or why not?

EUREKA
MATH

e.   Chris thought the median was 82. Do you agree? Why or why not?

5.   Calculate the mean, and compare it to the median. What do you observe about the two values? If the mean and median are both measures of center, why do you think one of them is smaller than the other?

Exercises 6–8: Finding Medians from Frequency Tables

6.   A third restaurant (Restaurant C) tallied the number of fries for a sample of bags of french fries and found the results below.

Number of Fries	Frequency
75	||
76	|
77	||
78	|||
79	卌
80	||||
81	|
82	|
83	
84	|||
85	|||
86	|

a.   How many bags of fries did they count?

b. What is the median number of fries for the sample of bags from this restaurant? Describe how you found your answer.

7. Robere wanted to look more closely at the data for bags of fries that contained a smaller number of fries and bags that contained a larger number of fries. He decided to divide the data into two parts. He first found the median of the whole data set and then divided the data set into the bottom half (the values in the ordered list that are before the median) and the top half (the values in the ordered list that are after the median).

a. List the 13 values in the bottom half. Find the median of these 13 values.

b. List the 13 values of the top half. Find the median of these 13 values.

8. Which of the three restaurants seems most likely to really have 82 fries in a typical bag? Explain your thinking.

Lesson Summary

The **median** is the middle value (or the mean of the two middle values) in a data set that has been ordered from smallest to largest. The median separates the data into two parts with the same number of data values below the median as above the median in the ordered list. To find a median, you first have to order the data. For an even number of data values, you find the average of the two middle numbers. For an odd number of data values, you use the middle value.

Problem Set

1. The amount of precipitation in each of the western states in the United States is given in the table as well as the dot plot.

State	Amount of Precipitation (inches)
WA	38.4
OR	27.4
CA	22.2
MT	15.3
ID	18.9
WY	12.9
NV	9.5
UT	12.2
CO	15.9
AZ	13.6
NM	14.6
AK	58.3
HI	63.7

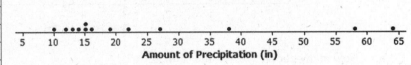

Source: http://www.currentresults.com/Weather/US/average-annual-state-precipitation.php

a. How do the amounts vary across the states?

b. Find the median. What does the median tell you about the amount of precipitation?

c. Do you think the mean or median would be a better description of the typical amount of precipitation? Explain your thinking.

---

2. Identify the following as true or false. If a statement is false, give an example showing why.

   a. The median is always equal to one of the values in the data set.

   b. The median is halfway between the least and greatest values in the data set.

   c. At most, half of the values in a data set have values less than the median.

   d. In a data set with 25 different values, if you change the two smallest values in the data set to smaller values, the median will not be changed.

   e. If you add 10 to every value in a data set, the median will not change.

3. Make up a data set such that the following is true:

   a. The data set has 11 different values, and the median is 5.

   b. The data set has 10 values, and the median is 25.

   c. The data set has 7 values, and the median is the same as the least value.

4. The dot plot shows the number of landline phones that a sample of people have in their homes.

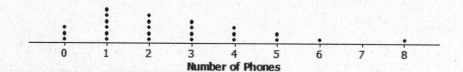

   a. How many people were in the sample?

   b. Why do you think three people have no landline phones in their homes?

   c. Find the median number of phones for the people in the sample.

5. The salaries of the Los Angeles Lakers for the 2012–2013 basketball season are given below. The salaries in the table are ordered from largest to smallest.

Player	Salary
Kobe Bryant	$27,849,149
Dwight Howard	$19,536,360
Pau Gasol	$19,000,000
Steve Nash	$8,700,000
Metta World Peace	$7,258,960
Steve Blake	$4,000,000
Jordan Hill	$3,563,600
Chris Duhon	$3,500,000
Jodie Meeks	$1,500,000
Earl Clark	$1,240,000
Devin Ebanks	$1,054,389
Darius Morris	$962,195
Antawn Jamison	$854,389
Robert Sacre	$473,604
Darius Johnson-Odom	$203,371

Source: www.basketball-reference.com/contracts/LAL.html

a. Just looking at the data, what do you notice about the salaries?

b. Find the median salary, and explain what it tells you about the salaries.

c. Find the median of the lower half of the salaries and the median of the upper half of the salaries.

d. Find the width of each of the following intervals. What do you notice about the size of the interval widths, and what does that tell you about the salaries?

    i.   Minimum salary to the median of the lower half:

    ii.   Median of the lower half to the median of the whole data set:

    iii. Median of the whole data set to the median of the upper half:

    iv.   Median of the upper half to the highest salary:

6. Use the salary table from above to answer the following.

a. If you were to find the mean salary, how do you think it would compare to the median? Explain your reasoning.

b. Which measure do you think would give a better picture of a typical salary for the Lakers, the mean or the median? Explain your thinking.

This page intentionally left blank

Lesson 13: Describing Variability Using the Interquartile Range (IQR)

Classwork

In Lesson 12, the median was used to describe a typical value for a data set. But the values in a data set vary around the median. What is a good way to indicate how the data vary when we use a median as an indication of a typical value? These questions are explored in the following exercises.

Exercises 1–4: More French Fries

1. In Lesson 12, you thought about the claim made by a chain restaurant that the typical number of french fries in a large bag was 82. Then, you looked at data on the number of fries in a bag from three of the restaurants.

   a. How do you think the data were collected, and what problems might have come up in collecting the data?

   b. What scenario(s) would give counts that might not be representative of typical bags?

2. The medians of the top half and the medians of the bottom half of the data for each of the three restaurants are as follows: Restaurant A—87.5 and 77; Restaurant B—83 and 76; Restaurant C—84 and 78. The difference between the medians of the two halves is called the *interquartile range,* or IQR.

   a. What is the IQR for each of the three restaurants?

   b. Which of the restaurants had the smallest IQR, and what does that tell you?

©2015 Great Minds eureka-math.org
G6-M5M6-SE-B3-1.3.1-01.2016

c.   The median of the bottom half of the data is called the *lower quartile* (denoted by Q1), and the median of the top half of the data is called the *upper quartile* (denoted by Q3). About what fraction of the data would be between the lower and upper quartiles? Explain your thinking.

3.   Why do you think that the median of the top half of the data is called the *upper quartile* and the median of the bottom half of the data is called the *lower quartile*?

4.

a.   Mark the quartiles for each restaurant on the graphs below.

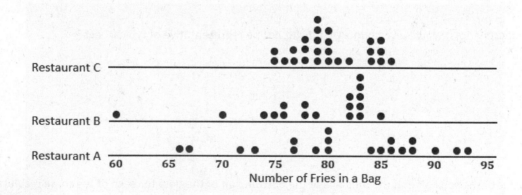

b.   Does the IQR help you decide which of the three restaurants seems most likely to really have 82 fries in a typical large bag? Explain your thinking.

©2015 Great Minds eureka-math.org
G6-M5M6-SE-B3-1.3.1-01.2016

Example 1: Finding the IQR

Read through the following steps. If something does not make sense to you, make a note, and raise it during class discussion. Consider the data: $1, 1, 3, 4, 6, 6, 7, 8, 10, 11, 11, 12, 15, 15, 17, 17, 17$

Creating an IQR:

a.   Put the data in order from smallest to largest.

b.   Find the minimum and maximum.

c.   Find the median.

d.   Find the lower quartile and upper quartile.

e.   Calculate the IQR by finding the difference between Q3 and Q1.

Exercise 5: When Should You Use the IQR?

5.   When should you use the IQR? The data for the 2012 salaries for the Lakers basketball team are given in the two
     plots below. (See Problem 5 in the Problem Set from Lesson 12.)

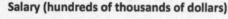

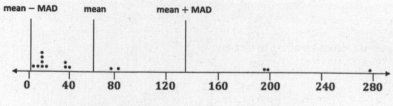

Salary (hundreds of thousands of dollars)

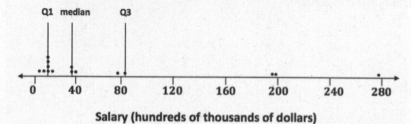

Salary (hundreds of thousands of dollars)

   a.   The data are given in hundreds of thousands of dollars. What would a salary of 40 hundred thousand dollars
        be?

   b.   The vertical lines on the top plot show the mean and the mean plus and minus the MAD. The bottom plot
        shows the median and the IQR. Which interval is a better picture of the typical salaries? Explain your thinking.

©2015 Great Minds eureka-math.org
G6-M5M6-SE-B3-1.3.1-01.2016

EUREKA
MATH

Exercise 6: On Your Own with IQRs

6. Create three different examples where you might collect data and where that data might have an IQR of 20. Define a median in the context of each example. Be specific about how the data might have been collected and the units involved. Be ready to describe what the median and IQR mean in each context.

   a.

   b.

   c.

Lesson Summary

To find the IQR, you order the data, find the median of the data, and then find the median of the bottom half of the data (the lower quartile) and the median of the top half of the data (the upper quartile). The IQR is the difference between the upper quartile and the lower quartile, which is the length of the interval that includes the middle half of the data. The median and the two quartiles divide the data into four sections, with about $\frac{1}{4}$ of the data in each section. Two of the sections are between the quartiles, so the interval between the quartiles would contain about 50% of the data.

Problem Set

1. The average monthly high temperatures (in degrees Fahrenheit) for St. Louis and San Francisco are given in the table below.

	Jan.	Feb.	Mar.	Apr.	May	June	July	Aug.	Sept.	Oct.	Nov.	Dec.
St. Louis	40	45	55	67	77	85	89	88	81	69	56	43
San Francisco	57	60	62	63	64	67	67	68	70	69	63	57

Data Source: http://www.weather.com

a. How do you think the data might have been collected?

b. Do you think it would be possible for $\frac{1}{4}$ of the temperatures in the month of July for St. Louis to be 95°F or above? Why or why not?

c. Make a prediction about how the values of the IQR for the temperatures for each city compare. Explain your thinking.

d. Find the IQR for the average monthly high temperature for each city. How do the results compare to what you predicted?

EUREKA MATH

2. The plot below shows the years in which each of 100 pennies were made.

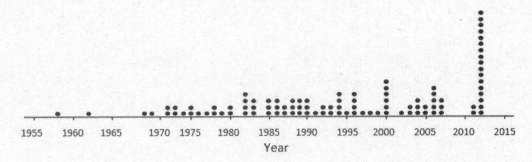

    a. What does the stack of 17 dots at 2012 representing 17 pennies tell you about the age of these pennies in 2014?

    b. Here is some information about the sample of 100 pennies. The mean year they were made is 1994; the first year any of the pennies were made was 1958; the newest pennies were made in 2012; Q1 is 1984, the median is 1994, and Q3 is 2006; the MAD is 11.5 years. Use the information to indicate the years in which the middle half of the pennies was made.

3. In each of parts (a)–(c), create a data set with at least 6 values such that it has the following properties:

    a. A small IQR and a big range (maximum – minimum)

    b. An IQR equal to the range

    c. The lower quartile is the same as the median.

4. Rank the following three data sets by the value of the IQR.

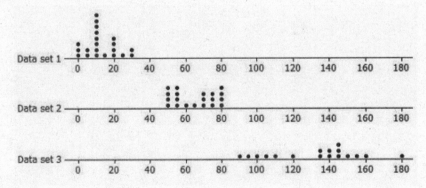

5. Here are the number of fries in each of the bags from Restaurant A:

    80, 72, 77, 80, 90, 85, 93, 79, 84, 73, 87, 67, 80, 86, 92, 88, 86, 88, 66, 77

    a. Suppose one bag of fries had been overlooked and that bag had only 50 fries. If that value is added to the data set, would the IQR change? Explain your reasoning.

    b. Will adding another data value always change the IQR? Give an example to support your answer.

This page intentionally left blank

Lesson 14: Summarizing a Distribution Using a Box Plot

Classwork

A box plot is a graph that is used to summarize a data distribution. What does the box plot tell us about the data distribution? How does the box plot indicate the variability of the data distribution? These questions are explored in this lesson.

Example 1: Time to Get to School

Consider the statistical question, "What is the typical amount of time it takes for a person in your class to get to school?" The amount of time it takes to get to school in the morning varies for the students in your class. Take a minute to answer the following questions. Your class will use this information to create a dot plot.

Write your name and an estimate of the number of minutes it took you to get to school today on a sticky note.

What were some of the things you had to think about when you made your estimate?

Exercises 1–4

Here is a dot plot of the estimates of the times it took students in Mr. S's class to get to school one morning.

Mr. S's Class

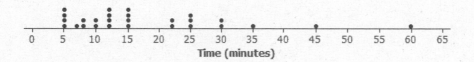

Time (minutes)

1. Put a line on the dot plot that you think separates the times into two groups—one group representing the longer times and the other group representing the shorter times.

2.   Put another line on the dot plot that separates out the times for students who live really close to the school. Add another line that separates out the times for students who take a very long time to get to school.

3.   Your dot plot should now be divided into four sections. Record the number of data values in each of the four sections.

4.   Share your marked-up dot plot with some of your classmates. Compare how each of you divided the dot plot into four sections.

Exercises 5–7: Time to Get to School

The times (in minutes) for the students in Mr. S's class have been put in order from smallest to largest and are shown below.

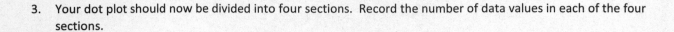

5 5 5 5 7 8 8 10 10 12 12 12 12 15 15 15 15 22 22 25 25 25 30 30 35 45 60

5.   What is the value of the median time to get to school for students in Mr. S's class?

6.   What is the value of the lower quartile? The upper quartile?

7. The lines on the dot plot below indicate the location of the median, the lower quartile, and the upper quartile. These lines divide the data set into four parts. About what fraction of the data values are in each part?

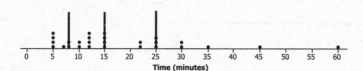

Mr. S's Class

Time (minutes)

Example 2: Making a Box Plot

A box plot is a graph made using the following five numbers: the smallest value in the data set, the lower quartile, the median, the upper quartile, and the largest value in the data set.

To make a box plot:

- Find the median of all of the data.
- Find Q1, the median of the bottom half of the data, and Q3, the median of the top half of the data.
- Draw a number line, and then draw a box that goes from Q1 to Q3.
- Draw a vertical line in the box at the value of the median.
- Draw a line segment connecting the minimum value to the box and a line segment that connects the maximum value to the box.

You will end up with a graph that looks something like this:

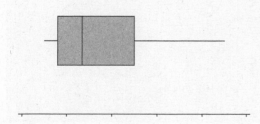

Now, use the given number line to make a box plot of the data below.

20, 21, 25, 31, 35, 38, 40, 42, 44

15   20   25   30   35   40   45

The five-number summary is as follows:

Min =

Q1 =

Median =

Q3 =

Max =

Exercises 8–11: A Human Box Plot

Consider again the sticky note that you used to write down the number of minutes it takes you to get to school. If possible, you and your classmates will form a human box plot of the number of minutes it takes students in your class to get to school.

8. Find the median of the group. Does someone represent the median? If not, who is the closest to the median?

9. Find the maximum and minimum of the group. Who are they?

10. Find Q1 and Q3 of the group. Does anyone represent Q1 or Q3? If not, who is the closest to Q1? Who is the closest to Q3?

11. Sketch the box plot for this data set.

EUREKA MATH

Lesson Summary

You learned how to make a box plot by doing the following:

- Finding the median of the entire data set.
- Finding Q1, the median of the bottom half of the data, and Q3, the median of the top half of the data.
- Drawing a number line and then drawing a box that goes from Q1 to Q3.
- Drawing a vertical line in the box at the value of the median.
- Drawing a line segment connecting the minimum value to the box and one that connects the maximum value to the box.

Problem Set

1. Dot plots for the amount of time it took students in Mr. S's and Ms. J's classes to get to school are below.

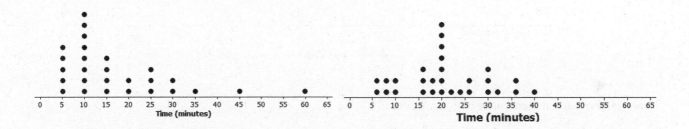

a. Make a box plot of the times for each class.

b. What is one thing you can see in the dot plot that you cannot see in the box plot? What is something that is easier to see in the box plot than in the dot plot?

2. The dot plot below shows the vertical jump of some NBA players. A vertical jump is how high a player can jump from a standstill. Draw a box plot of the heights for the vertical jumps of the NBA players above the dot plot.

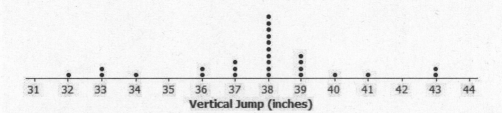

Lesson 14:   Summarizing a Distribution Using a Box Plot

S.105

©2015 Great Minds eureka-math.org
G6-M5M6-SE-B3-1.3.1-01.2016

3. The mean daily temperatures in degrees Fahrenheit for the month of February for a certain city are as follows:

    4, 11, 14, 15, 17, 20, 30, 23, 20, 35, 35, 31, 34, 23, 15, 19, 39, 22, 15, 15, 19, 39, 22, 23, 29, 26, 29, 29

    a. Make a box plot of the temperatures.

    b. Make a prediction about the part of the United States you think the city might be located. Explain your reasoning.

    c. Describe the temperature data distribution. Include a description of center and spread.

4. The box plot below summarizes data from a survey of households about the number of dogs they have. Identify each of the following statements as true or false. Explain your reasoning in each case.

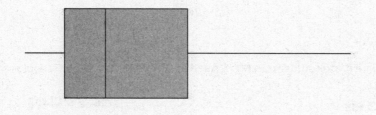

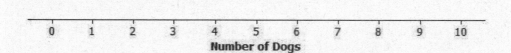

Number of Dogs

    a. The maximum number of dogs per house is 8.

    b. At least $\frac{1}{2}$ of the houses have 2 or more dogs.

    c. All of the houses have dogs.

    d. Half of the houses surveyed have between 2 and 4 dogs.

    e. Most of the houses surveyed have no dogs.

EUREKA
MATH™

Lesson 15: More Practice with Box Plots

Classwork

You reach into a jar of Tootsie Pops. How many Tootsie Pops do you think you could hold in one hand? Do you think the number you could hold is greater than or less than what other students can hold? Is the number you could hold a typical number of Tootsie Pops? This lesson examines these questions.

Example 1: Tootsie Pops

Ninety-four people were asked to grab as many Tootsie Pops as they could hold. Here is a box plot for these data. Are you surprised?

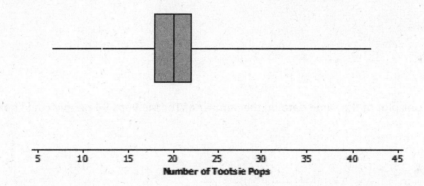

Number of Tootsie Pops

Exercises 1–5

1. What might explain the variability in the number of Tootsie Pops that the 94 people were able to hold?

2. Use a box plot to estimate the values in the five-number summary.

3. Describe how the box plot can help you understand differences in the numbers of Tootsie Pops people could hold.

4. Here is Jayne's description of what she sees in the box plot. Do you agree or disagree with her description? Explain your reasoning.

    "One person could hold as many as 42 Tootsie Pops. The number of Tootsie Pops people could hold was really different and spread about equally from 7 to 42. About one-half of the people could hold more than 20 Tootsie Pops."

5. Here is a different box plot of the same data on the number of Tootsie Pops 94 people could hold.

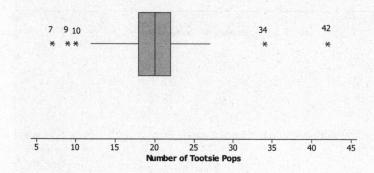

    a. Why do you suppose there are five values that are shown as separate points and are labeled?

    b. Does knowing these data values change anything about your responses to Exercises 1 to 4 above?

EUREKA
MATH™

Exercises 6–10: Maximum Speeds

The maximum speeds of selected birds and land animals are given in the tables below.

Bird	Speed (mph)
Peregrine falcon	242
Swift bird	120
Spine-tailed swift	106
White-throated needle tail	105
Eurasian hobby	100
Pigeon	100
Frigate bird	95
Spur-winged goose	88
Red-breasted merganser	80
Canvasback duck	72
Anna's hummingbird	61.06
Ostrich	60

Land Animal	Speed (mph)
Cheetah	75
Free-tailed bat (in flight)	60
Pronghorn antelope	55
Lion	50
Wildebeest	50
Jackrabbit	44
African wild dog	44
Kangaroo	45
Horse	43.97
Thomson's gazelle	43
Greyhound	43
Coyote	40
Mule deer	35
Grizzly bear	30
Cat	30
Elephant	25
Pig	9

Data sources: *Natural History Magazine*, March 1974, copyright 1974; The American Museum of Natural History; and James G. Doherty, general curator, The Wildlife Conservation Society; http://www.thetravelalmanac.com/lists/animals-speed.htm; http://en.wikipedia.org/wiki/Fastest_animals

6. As you look at the speeds, what strikes you as interesting?

7. Do birds or land animals seem to have the greatest variability in speeds? Explain your reasoning.

8. Find the five-number summary for the speeds in each data set. What do the five-number summaries tell you about the distribution of speeds for each data set?

9. Use the five-number summaries to make a box plot for each of the two data sets.

```
|----|----|----|----|----|----|----|----|----|----|----|
0    25   50   75   100  125  150  175  200  225  250
                 Maximum Speed of Birds (mph)
```

```
|----|----|----|----|----|----|----|----|----|----|----|
0    25   50   75   100  125  150  175  200  225  250
              Maximum Speed of Land Animals (mph)
```

10. Write several sentences describing the speeds of birds and land animals.

EUREKA
MATH™

Exercises 11–15: What Is the Same, and What Is Different?

Consider the following box plots, which show the number of correctly answered questions on a 20-question quiz for students in three different classes.

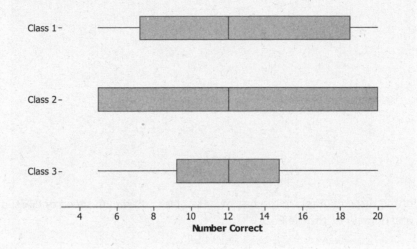

11. Describe the variability in the scores of each of the three classes.

12.

    a.   Estimate the interquartile range for each of the three sets of scores.

    b.   What fraction of students would have scores in the interval that extends from the lower quartile to the upper quartile?

    c.   What does the value of the IQR tell you about how the scores are distributed?

13. Which class do you believe performed the best? Be sure to use information from the box plots to back up your answer.

14.

    a. Find the IQR for the three data sets in the first two examples: maximum speed of birds, maximum speed of land animals, and number of Tootsie Pops.

    b. Which data set had the highest percentage of data values between the lower quartile and the upper quartile? Explain your thinking.

15. A teacher asked students to draw a box plot with a minimum value at 34 and a maximum value at 64 that had an interquartile range of 10. Jeremy said he could not draw just one because he did not know where to put the box on the number line. Do you agree with Jeremy? Why or why not?

©2015 Great Minds eureka-math.org
G6-M5M6-SE-B3-1.3.1-01.2016

Problem Set

1. The box plot below summarizes the maximum speeds of certain kinds of fish.

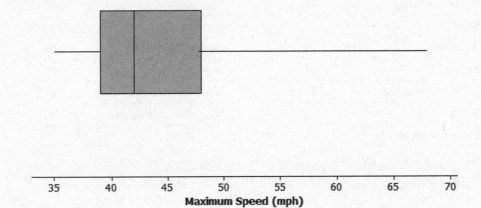

Maximum Speed (mph)

  a. Estimate the values in the five-number summary from the box plot.

  b. The fastest fish is the sailfish at 68 mph, followed by the marlin at 50 mph. What does this tell you about the spread of the fish speeds in the top quarter of the box plot?

  c. Use the five-number summary and the IQR to describe the speeds of the fish.

2. Suppose the interquartile range for the number of hours students spent playing video games during the school week was 10. What do you think about each of the following statements? Explain your reasoning.

  a. About half of the students played video games for 10 hours during a school week.

  b. All of the students played at least 10 hours of video games during the school week.

  c. About half of the class could have played video games from 10 to 20 hours a week or from 15 to 25 hours.

3. Suppose you know the following for a data set: The minimum value is 130, the lower quartile is 142, the IQR is 30, half of the data are less than 168, and the maximum value is 195.

  a. Think of a context for which these numbers might make sense.

  b. Sketch a box plot.

  c. Are there more data values above or below the median? Explain your reasoning.

4. The speeds for the fastest dogs are given in the table below.

Breed	Speed (mph)
Greyhound	45
African wild dog	44
Saluki	43
Whippet	36
Basanji	35
German shepherd	32
Vizsla	32
Doberman pinscher	30

Breed	Speed (mph)
Irish wolfhound	30
Dalmatian	30
Border collie	30
Alaskan husky	28
Giant schnauzer	28
Jack Russell terrier	25
Australian cattle dog	20

Data source: http://www.vetstreet.com/our-pet-experts/meet-eight-of-the-fastest-dogs-on-the-planet; http://canidaepetfood.blogspot.com/2012/08/which-dog-breeds-are-fastest.html

a. Find the five-number summary for this data set, and use it to create a box plot of the speeds.

b. Why is the median not in the center of the box?

c. Write a few sentences telling your friend about the speeds of the fastest dogs.

EUREKA MATH

Lesson 16: Understanding Box Plots

Classwork

Exercise 1: Supreme Court Chief Justices

1. The Supreme Court is the highest court of law in the United States, and it makes decisions that affect the whole country. The chief justice is appointed to the court and is a justice the rest of his life unless he resigns or becomes ill. Some people think that this means that the chief justice serves for a very long time. The first chief justice was appointed in 1789.

   The table shows the years in office for each of the chief justices of the Supreme Court as of 2013:

Name	Number of Years	Year Appointed
John Jay	6	1789
John Rutledge	1	1795
Oliver Ellsworth	4	1796
John Marshall	34	1801
Roger Brooke Taney	28	1836
Salmon P. Chase	9	1864
Morrison R. Waite	14	1874
Melville W. Fuller	22	1888
Edward D. White	11	1910
William Howard Taft	9	1921
Charles Evens Hughes	11	1930
Harlan Fiske Stone	5	1941
Fred M. Vinson	7	1946
Earl Warren	16	1953
Warren E. Burger	17	1969
William H. Rehnquist	19	1986
John G. Roberts	8	2005

   Data source: http://en.wikipedia.org/wiki/List_of_Justices_of_the_Supreme_Court_of_the_United_States

   Use the table to answer the following:

   a. Which chief justice served the longest term, and which served the shortest term? How many years did each of these chief justices serve?

©2015 Great Minds eureka-math.org
G6-M5M6-SE-B3-1.3.1-01.2016

b.   What is the median number of years these chief justices have served on the Supreme Court? Explain how you found the median and what it means in terms of the data.

c.   Make a box plot of the years the justices served. Describe the shape of the distribution and how the median and IQR relate to the box plot.

d.   Is the median halfway between the least and the most number of years served? Why or why not?

Exercises 2–3: Downloading Songs

2.   A broadband company timed how long it took to download 232 four-minute songs on a dial-up connection. The dot plot below shows their results.

a.   What can you observe about the download times from the dot plot?

b.   Is it easy to tell whether or not 12.5 minutes is in the top quarter of the download times?

EUREKA
MATH

c.   The box plot of the data is shown below. Now, answer parts (a) and (b) above using the box plot.

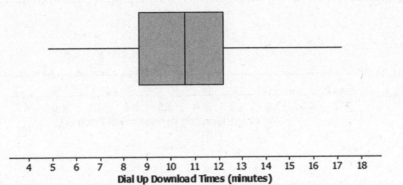

d.   What are the advantages of using a box plot to summarize a large data set? What are the disadvantages?

3.   Molly presented the box plots below to argue that using a dial-up connection would be better than using a broadband connection. She argued that the dial-up connection seems to have less variability around the median even though the overall range seems to be about the same for the download times using broadband. What would you say?

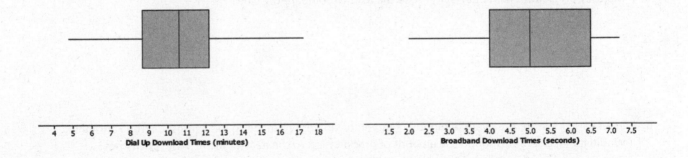

EUREKA
MATH™

Lesson 16:   Understanding Box Plots

S.117

©2015 Great Minds eureka-math.org
G6-M5M6-SE-B3-1.3.1-01.2016

Exercises 4–5: Rainfall

4. Data on the average rainfall for each of the twelve months of the year were used to construct the two dot plots below.

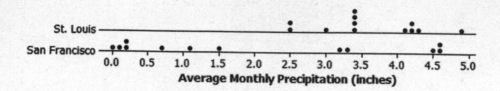

a. How many data points are in each dot plot? What does each data point represent?

b. Make a conjecture about which city has the most variability in the average monthly amount of precipitation and how this would be reflected in the IQRs for the data from both cities.

c. Based on the dot plots, what are the approximate values of the interquartile ranges (IQRs) for the average monthly precipitations for each city? Use the IQRs to compare the cities.

d. In an earlier lesson, the average monthly temperatures were rounded to the nearest degree Fahrenheit. Would it make sense to round the amount of precipitation to the nearest inch? Why or why not?

EUREKA
MATH

5. Use the data from Exercise 4 to answer the following.

   a. Make a box plot of the monthly precipitation amounts for each city using the same scale.

```
   0.0    0.5    1.0    1.5    2.0    2.5    3.0    3.5    4.0    4.5    5.0

              Average Monthly Precipitation in St. Louis (inches)
```

```
   0.0    0.5    1.0    1.5    2.0    2.5    3.0    3.5    4.0    4.5    5.0

           Average Monthly Precipitation in San Francisco (inches)
```

   b. Compare the percent of months that have above 2 inches of precipitation for the two cities. Explain your thinking.

   c. How does the top 25% of the average monthly precipitations compare for the two cities?

   d. Describe the intervals that contain the smallest 25% of the average monthly precipitation amounts for each city.

e.   Think about the dot plots and the box plots. Which representation do you think helps you the most in understanding how the data vary?

Note: The data used in this problem are displayed in the table below.

Average Precipitation (inches)

	Jan.	Feb.	Mar.	Apr.	May	June	July	Aug.	Sept.	Oct.	Nov.	Dec.
St. Louis	2.45	2.48	3.36	4.10	4.80	4.34	4.19	3.41	3.38	3.43	4.22	2.96
San Francisco	4.5	4.61	3.76	1.46	0.70	0.16	0	0.06	0.21	1.12	3.16	4.56

Data source: http://www.weather.com

Problem Set

1. The box plots below summarize the ages at the time of the award for leading actress and leading actor Academy Award winners.

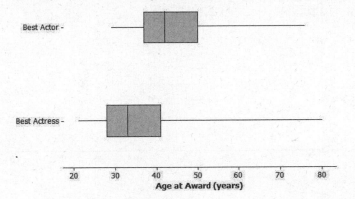

Data source: http://en.wikipedia.org/wiki/List_of_Best_Actor_winners_by_age_at_win

http://en.wikipedia.org/wiki/List_of_Best_Actress_winners_by_age_at_win

a. Based on the box plots, do you think it is harder for an older woman to win an Academy Award for best actress than it is for an older man to win a best actor award? Why or why not?

b. The oldest female to win an Academy Award was Jessica Tandy in 1990 for *Driving Miss Daisy*. The oldest actor was Henry Fonda for *On Golden Pond* in 1982. How old were they when they won the award? How can you tell? Were they a lot older than most of the other winners?

c. The 2013 winning actor was Daniel Day-Lewis for *Lincoln*. He was 55 years old at that time. What can you say about the percent of male award winners who were older than Daniel Day-Lewis when they won their Oscars?

d. Use the information provided by the box plots to write a paragraph supporting or refuting the claim that fewer older actresses than actors win Academy Awards.

2. The scores of sixth and seventh graders on a test about polygons and their characteristics are summarized in the box plots below.

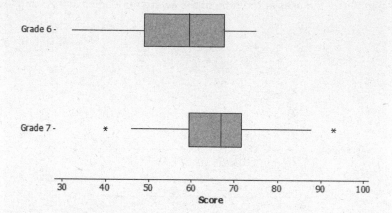

a. In which grade did the students do the best? Explain how you can tell.

b. Why do you think two of the data values for Grade 7 are not part of the line segments?

c. How do the median scores for the two grades compare? Is this surprising? Why or why not?

d. How do the IQRs compare for the two grades?

3. A formula for the IQR could be written as $Q3 - Q1 = IQR$. Suppose you knew the IQR and the Q1. How could you find the Q3?

4. Consider the statement, "Historically, the average length of service as chief justice on the Supreme Court has been less than 15 years; however, since 1969 the average length of service has increased." Use the data given in Exercise 1 to answer the following questions.

a. Do you agree or disagree with the statement? Explain your thinking.

b. Would your answer change if you used the median number of years rather than the mean?

Lesson 17: Developing a Statistical Project

Classwork

Exploratory Challenge

Review of Statistical Questions

Statistical questions you investigated in this module included the following:

- How many hours of sleep do sixth graders typically get on a night when there is school the next day?
- What is the typical number of books read over the course of 6 months by a sixth grader?
- What is the typical heart rate of a student in a sixth-grade class?
- How many hours does a sixth grader typically spend playing a sport or a game outdoors?
- What are the head circumferences of adults interested in buying baseball hats?
- How long is the battery life of a certain brand of batteries?
- How many pets do students have?
- How long does it take students to get to school?
- What is a typical daily temperature in New York City?
- What is the typical weight of a backpack for students at a certain school?
- What is the typical number of french fries in a large order from a fast food restaurant?
- What is the typical number of minutes a student spends on homework each day?
- What is the typical height of a vertical jump for a player in the NBA?

What do these questions have in common?

Why do several of these questions include the word *typical*?

A Review of a Statistical Investigation

Recall from the very first lesson in this module that a statistical question is a question answered by data that you anticipate will vary.

Let's review the steps of a statistical investigation.

> Step 1:   Pose a question that can be answered by data.
>
> Step 2:   Collect appropriate data.
>
> Step 3:   Summarize the data with graphs and numerical summaries.
>
> Step 4:   Answer the question posed in Step 1 using the numerical summaries and graphs.

The first step is to pose a statistical question. Select one of the questions investigated in this module, and write it in the following Statistical Study Review Template.

The second step is to collect the data. In all of these investigations, you were given data. How do you think the data for the question you selected in Step 1 were collected? Write your answer in the summary below for Step 2.

The third step involves the various ways you summarize the data. List the various ways you summarized the data in the space for Step 3.

Statistical Study Review Template

Step 1: Pose a statistical question.
Step 2: Collect the data.
Step 3: Summarize the data.

Step 4: Answer the question.

Developing Statistical Questions

Now it is your turn to answer a statistical question based on data you collect. Before you collect the data, explore possible statistical questions. For each question, indicate the data that you would collect and summarize to answer the question. Also, indicate how you plan to collect the data.

Think of questions that could be answered by data collected from members of your class or school or data that could be collected from recognized websites (such as the American Statistical Association and the Census at School project). Your teacher will need to approve both your question and your plan to collect data before data are collected.

As a class, explore possibilities for a statistical investigation. Record some of the ideas discussed by your class using the following table.

Possible Statistical Questions	What data would be collected, and how would the data be collected?

After discussing several of the possibilities for a statistical project, prepare a statistical question and a plan to collect the data. After your teacher approves your question and data collection plan, begin collecting the data. Carefully organize your data as you begin developing the numerical and graphical summaries to answer your statistical question. In future lessons, you will be directed to begin creating a poster or an outline of a presentation that will be shared with your teacher and other members of your class.

Complete the following to present to your teacher:

1. The statistical question for my investigation is:

2. Here is how I propose to collect my data. (Include how you are going to collect your data and a clear description of what you plan to measure or count.)

Lesson Summary

A statistical investigation involves a four-step investigative process:

- Pose questions that can be answered by data.
- Design a plan for collecting appropriate data, and then use the plan to collect data.
- Analyze the data.
- Interpret results and draw valid conclusions from the data to answer the question posed.

Problem Set

Your teacher will outline steps you are expected to complete in the next several days to develop this project. Keep in mind that the first step is to formulate a statistical question. With one of the statistical questions posed in this lesson or with a new one developed in this lesson, describe your question and plan to collect and summarize data. Complete the process as outlined by your teacher.

This page intentionally left blank

Lesson 18: Connecting Graphical Representations and Numerical Summaries

Classwork

Here is a data set of the ages (in years) of 43 participants who ran in a 5-kilometer race.

20	30	30	35	36	34	38	46
45	18	43	23	47	27	21	30
32	32	31	32	36	74	41	41
51	61	50	34	34	34	35	28
57	26	29	49	41	36	37	41
38	30	30					

Here are some summary statistics, a dot plot, and a histogram for the data:

Minimum = 18, Q1 = 30, Median = 35, Q3 = 41, Maximum = 74; Mean = 36.8, MAD = 8.1

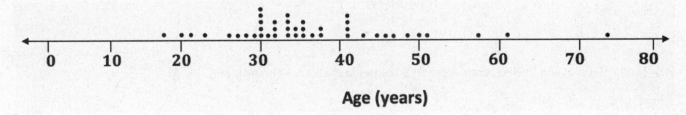

Age (years)

Histogram of Participant Ages in a 5K Race

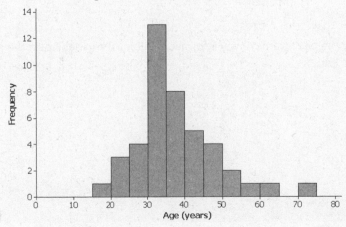

©2015 Great Minds eureka-math.org
G6-M5M6-SE-B3-1.3.1-01.2016

Exercises 1–7

1. Based on the histogram, would you describe the shape of the data distribution as approximately symmetric or as skewed? Would you have reached this same conclusion looking at the dot plot?

2. If there had been 500 participants instead of just 43, would you use a dot plot or a histogram to display the data?

3. What is something you can see in the dot plot that is not as easy to see in the histogram?

4. Do the dot plot and the histogram seem to be centered in about the same place?

5. Do both the dot plot and the histogram convey information about the variability in the age distribution?

EUREKA
MATH

6. If you did not have the original data set and only had the dot plot and the histogram, would you be able to find the value of the median age from the dot plot?

7. Explain why you would only be able to estimate the value of the median if you only had a histogram of the data.

Exercises 8–12: Graphs and Numerical Summaries

8. Suppose that a newspaper article was written about the race. The article included the histogram shown here and also said, "The race attracted many older runners this year. The median age was 45." Based on the histogram, how can you tell that this is an incorrect statement?

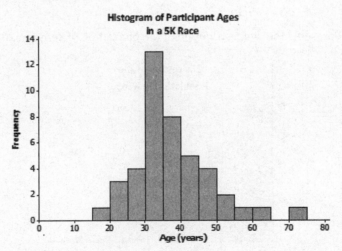

9. One of the histograms below is another correctly drawn histogram for the runners' ages. Select the correct histogram, and explain how you determined which graph is correct (and which one is incorrect) based on the summary measures and dot plot.

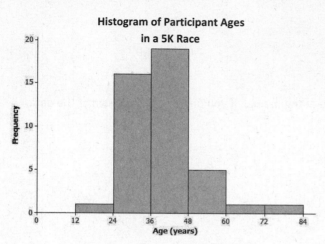

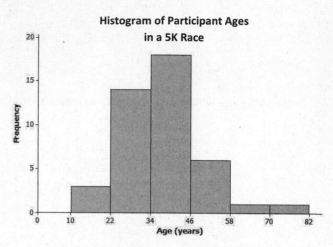

10. The histogram below represents the age distribution of the population of Kenya in 2010.

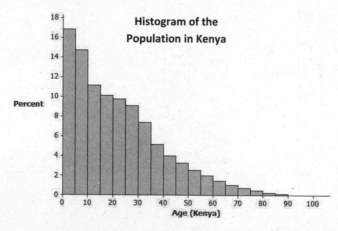

    a. How do we know from the graph above that the first quartile (Q1) of this age distribution is between 5 and 10 years of age?

EUREKA
MATH

b. Someone believes that the median age in Kenya is about 30. Based on the histogram, is 30 years a good estimate of the median age for Kenya? Explain why it is or why it is not.

11. The histogram below represents the age distribution of the population of the United States in 2010. Based on the histogram, which of the following ranges do you think includes the median age for the United States: 20–30, 30–40, or 40–50? Why?

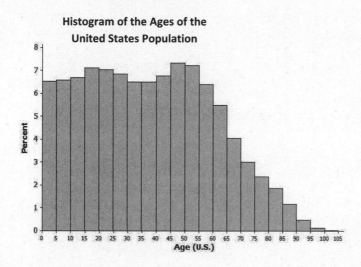

12. Consider the following three dot plots. Note: The same scale is used in each dot plot.

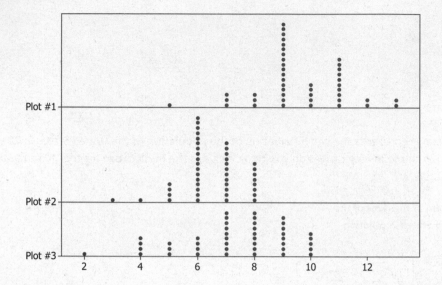

a. Which dot plot has a median of 8? Explain why you selected this dot plot over the other two.

b. Which dot plot has a mean of 9.6? Explain why you selected this dot plot over the other two.

c. Which dot plot has a median of 6 and a range of 5? Explain why you selected this dot plot over the other two.

EUREKA
MATH

©2015 Great Minds eureka-math.org
G6-M5M6-SE-B3-1.3.1-01.2016

Problem Set

1. The following histogram shows the amount of coal produced (by state) for the 20 largest coal-producing states in 2011. Many of these states produced less than 50 million tons of coal, but one state produced over 400 million tons (Wyoming). For the histogram, which *one* of the three sets of summary measures could match the graph? For each choice that you eliminate, give at least one reason for eliminating the choice.

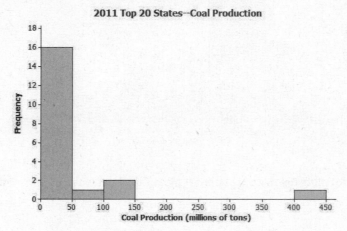

Source: U.S. Coal Production by State data as reported by the National Mining Association from http://www.nma.org/pdf/c_production_state_rank.pdf, accessed May 5, 2013

a. Minimum = 1, Q1 = 12, Median = 36, Q3 = 57, Maximum = 410; Mean = 33, MAD = 2.76

b. Minimum = 2, Q1 = 13.5, Median = 27.5, Q3 = 44, Maximum = 439; Mean = 54.6, MAD = 52.36

c. Minimum = 10, Q1 = 37.5, Median = 62, Q3 = 105, Maximum = 439; Mean = 54.6, MAD = 52.36

2. The heights (rounded to the nearest inch) of the 41 members of the 2012–2013 University of Texas Men's Swimming and Diving Team are shown in the dot plot below.

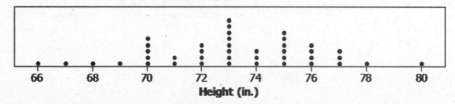

Source: http://www.texassports.com accessed April 30, 2013

a. Use the dot plot to determine the 5-number summary (minimum, lower quartile, median, upper quartile, and maximum) for the data set.

b. Based on this dot plot, make a histogram of the heights using the following intervals: 66 to < 68 inches, 68 to < 70 inches, and so on.

3.   Data on the weight (in pounds) of 143 wild bears are summarized in the histogram below.

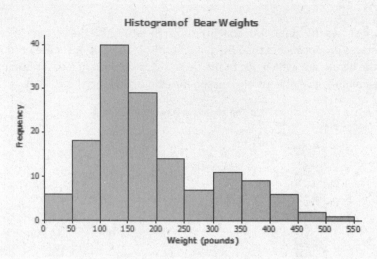

Which *one* of the three dot plots below could be a dot plot of the bear weight data? Explain how you determined which the correct plot is.

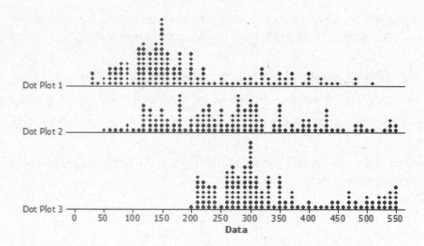

©2015 Great Minds eureka-math.org
G6-M5M6-SE-B3-1.3.1-01.2016

Lesson 19: Comparing Data Distributions

Classwork

Suppose that you are interested in comparing the weights of adult male polar bears and the weights of adult male grizzly bears. If data were available on the weights of these two types of bears, they could be used to answer questions such as:

   Do adult polar bears typically weigh less than adult grizzly bears?
   Are the weights of adult polar bears similar to each other, or do the weights tend to differ a lot from bear to bear?
   Are the weights of adult polar bears more consistent than the weights of adult grizzly bears?

These questions could be answered most easily by comparing the weight distributions for the two types of bears. Graphs of the data distributions (such as dot plots, box plots, or histograms) that are drawn side by side and that are drawn to the same scale make it easy to compare data distributions in terms of center, variability, and shape.

In this lesson, when two or more data distributions are presented, think about the following:

   How are the data distributions similar?
   How are the data distributions different?
   What do the similarities and differences tell you in the context of the data?

Example 1: Comparing Groups Using Box Plots

Recall that a *box plot* is a visual representation of a five-number summary. The box part of a box plot is drawn so that the width of the box represents the IQR. The distance from the far end of the line on the left to the far end of the line on the right represents the range.

If two box plots (each representing a different distribution) are drawn side by side using the same scale, it is easy to compare the values in the five-number summaries for the two distributions and to visually compare the IQRs and ranges.

Here is a data set of the ages of 43 participants in a 5-kilometer race (shown in a previous lesson).

20	30	30	35	36	34	38	46
45	18	43	23	47	27	21	30
32	32	31	32	36	74	41	41
51	61	50	34	34	34	35	28
57	26	29	49	41	36	37	41
38	30	30					

Here is the five-number summary for the data: Minimum = 18, Q1 = 30, Median = 35, Q3 = 41, Maximum = 74.

There was also a 15-kilometer race. The ages of the 55 participants in that race appear below.

47	19	30	30	36	37	35	39
19	49	47	16	45	22	50	27
19	20	30	32	32	31	32	37
22	81	43	43	54	66	53	35
22	35	35	36	28	61	26	29
38	52	43	37	38	43	39	30
58	30	48	49	54	56	58	

Does the longer race appear to attract different runners in terms of age? Below are side-by-side box plots that may help answer that question. Side-by-side box plots are two or more box plots drawn using the same scale. What do you notice about the two box plots?

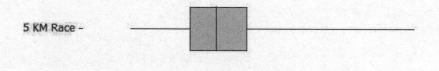
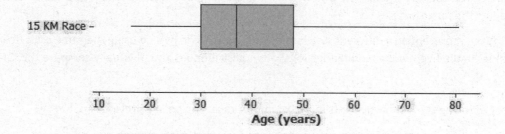

EUREKA MATH™

Exercises 1–6

1. Based on the box plots, estimate the values in the five-number summary for the age in the 15-kilometer race data set.

2. Do the two data sets have the same median? If not, which race had the higher median age?

3. Do the two data sets have the same IQR? If not, which distribution has the greater spread in the middle 50% of its distribution?

4. Which race had the smaller overall range of ages? What do you think the range of ages is for the 15-kilometer race?

5. Which race had the oldest runner? About how old was this runner?

6. Now, consider just the youngest 25% of the runners in the 15-kilometer race. How old was the youngest runner in this group? How old was the oldest runner in this group? How does that compare with the 5-kilometer race?

©2015 Great Minds eureka-math.org
G6-M5M6-SE-B3-1.3.1-01.2016

Exercises 7–12: Comparing Box Plots

In 2012, Major League Baseball had two leagues: an American League of 14 teams and a National League of 16 teams. Jesse wondered if American League teams have higher batting averages and on-base percentages. (Higher values are better.) Use the following box plots to investigate. (Source: http://mlb.mlb.com/stats/sortable.jsp, accessed May 13, 2013)

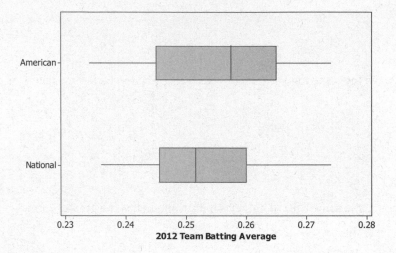

7. Was the highest American League team batting average very different from the highest National League team batting average? Approximately how large was the difference, and which league had the higher maximum value?

8. Was the range of American League team batting averages very different or only slightly different from the range of National League team batting averages?

9. Which league had the higher median team batting average? Given the scale of the graph and the range of the data sets, does the difference between the median values for the two leagues seem to be small or large? Explain why you think it is small or large.

Lesson 19:   Comparing Data Distributions

©2015 Great Minds eureka-math.org
G6-M5M6-SE-B3-1.3.1-01.2016

EUREKA MATH™

10. Based on the box plots below for on-base percentage, which three summary values (from the five-number summary) appear to be the same or virtually the same for both leagues?

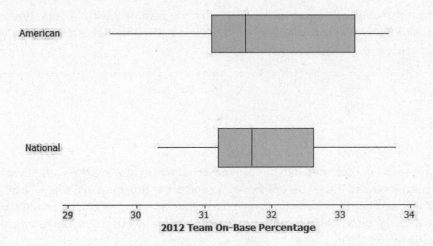

11. Which league's data set appears to have less variability? Explain.

12. Recall that Jesse wondered if American League teams have higher batting averages and on-base percentages. Based on the box plots given above, what would you tell Jesse?

Lesson Summary

When comparing the distribution of a quantitative variable for two or more distinct groups, it is useful to display the groups' distributions side by side using graphs drawn to the same scale. This makes it easier to describe the similarities and differences in the distributions of the groups.

Problem Set

1. College athletic programs are separated into divisions based on school size, available athletic scholarships, and other factors. A researcher wondered if members of swimming and diving programs in Division I (usually large schools that offer athletic scholarships) tend to be taller than the swimmers and divers in Division III programs (usually smaller schools that do not offer athletic scholarships). To begin the investigation, the researcher creates side-by-side box plots for the heights (in inches) of 41 male swimmers and divers at Mountain Vista University (a Division I program) and the heights (in inches) of 10 male swimmers and divers at Eaglecrest College (a Division III program).

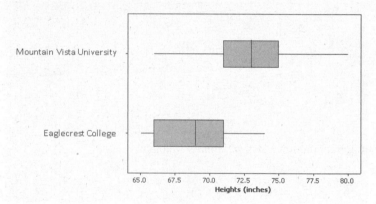

a. Which data set has the smaller range?

b. True or false: A swimmer who had a height equal to the median for the Mountain Vista University would be taller than the median height of swimmers and divers at Eaglecrest College.

c. To be thorough, the researcher will examine many other colleges' sports programs to further investigate the claim that members of swimming and diving programs in Division I are generally taller than the swimmers and divers in Division III. But given the graph above, in this initial stage of her research, do you think that the claim might be valid? Carefully support your answer using summary measures or graphical attributes.

©2015 Great Minds eureka-math.org
G6-M5M6-SE-B3-1.3.1-01.2016

2. Data on the weights (in pounds) of 100 polar bears and 50 grizzly bears are summarized in the box plots shown below.

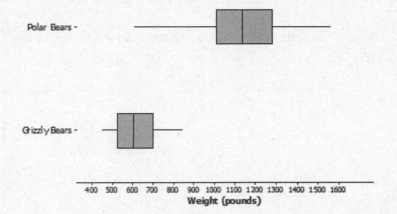

a. True or false: At least one of the polar bears weighed more than the heaviest grizzly bear. Explain how you know.

b. True or false: Weight differs more from bear to bear for polar bears than for grizzly bears. Explain how you know.

c. Which type of bear tends to weigh more? Explain.

3. Many movie studios rely heavily on viewer data to determine how a movie will be marketed and distributed. Recently, previews of a soon-to-be-released movie were shown to 300 people. Each person was asked to rate the movie on a scale of 0 to 10, with 10 representing "best movie I have ever seen" and 0 representing "worst movie I have ever seen."

Below are some side-by-side box plots that summarize the ratings by gender and by age.

For 150 women and 150 men:

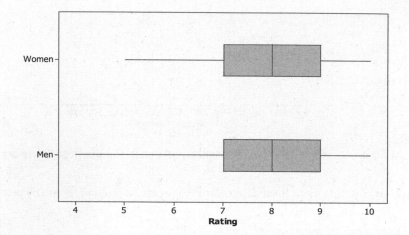

For 3 age groups:

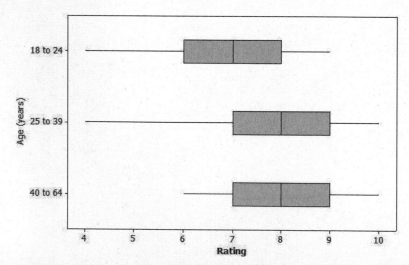

a. Does it appear that the men and women rated the film in a similar manner or in a very different manner? Write a few sentences explaining your answer using comparative information about center and variability.

b. It appears that the film tended to receive better ratings from the older members of the group. Write a few sentences using comparative measures of center and spread or aspects of the graphical displays to justify this claim.

EUREKA
MATH

Lesson 20: Describing Center, Variability, and Shape of a Data Distribution from a Graphic Representation

Classwork

Great Lakes yellow perch are fish that live in each of the five Great Lakes and many other lakes in the eastern and upper Great Lakes regions of the United States and Canada. Both countries are actively involved in efforts to maintain a healthy population of perch in these lakes.

Example 1: The Great Lakes Yellow Perch

Scientists collected data from many yellow perch because they were concerned about the survival of the yellow perch. What data do you think researchers might want to collect about these perch?

Scientists captured yellow perch from a lake in this region. They recorded data on each fish and then returned each fish to the lake. Consider the following histogram of data on the length (in centimeters) for a sample of yellow perch.

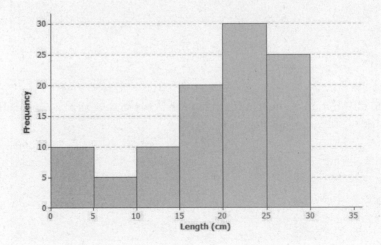

Exercises 1–11

Scientists were concerned about the survival of the yellow perch as they studied the histogram.

1. What statistical question could be answered based on this data distribution?

2. Use the histogram to complete the following table:

Length of Fish in Centimeters (cm)	Number of Fish
$0 \leq 5$ cm	
$5 \leq 10$ cm	
$10 \leq 15$ cm	
$15 \leq 20$ cm	
$20 \leq 25$ cm	
$25 \leq 30$ cm	

3. The length of each fish in the sample was measured and recorded before the fish was released back into the lake. How many yellow perch were measured in this sample?

4. Would you describe the distribution of the lengths of the fish in the sample as a skewed distribution or as an approximately symmetric distribution? Explain your answer.

©2015 Great Minds eureka-math.org
G6-M5M6-SE-B3-1.3.1-01.2016

5. What percentage of fish in the sample were less than 10 centimeters in length?

6. If the smallest fish in this sample was 2 centimeters in length, what is your estimate of an interval of lengths that would contain the lengths of the shortest 25% of the fish? Explain how you determined your answer.

7. If the length of the largest yellow perch was 29 centimeters, what is your estimate of an interval of lengths that would contain the lengths of the longest 25% of the fish?

8. Estimate the median length of the yellow perch in the sample. Explain how you determined your estimate.

9. Based on the shape of this data distribution, do you think the mean length of a yellow perch would be greater than, less than, or the same as your estimate of the median? Explain your answer.

10. Recall that the mean length is the balance point of the distribution of lengths. Estimate the mean length for this sample of yellow perch.

11. The length of a yellow perch is used to estimate the age of the fish. Yellow perch typically grow throughout their lives. Adult yellow perch have lengths between 10 and 30 centimeters. How many of the yellow perch in this sample would be considered adult yellow perch? What percentage of the fish in the sample are adult fish?

Example 2: What Would a Better Distribution Look Like?

Yellow perch are part of the food supply of larger fish and other wildlife in the Great Lakes region. Why do you think that the scientists worried when they saw the histogram of fish lengths given previously in Exercise 2.

Sketch a histogram representing a sample of 100 yellow perch lengths that you think would indicate the perch are not in danger of dying out.

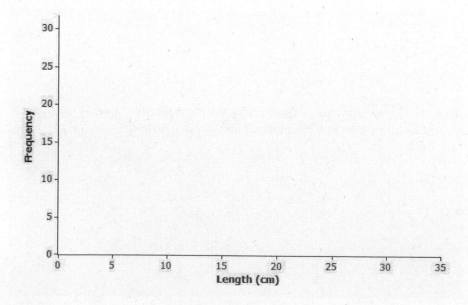

Lesson 20:   Describing Center, Variability, and Shape of a Data Distribution from a
                          Graphic Representation

©2015 Great Minds eureka-math.org
G6-M5M6-SE-B3-1.3.1-01.2016

Exercises 12–17: Estimating the Variability in Yellow Perch Lengths

You estimated the median length of yellow perch from the first sample in Exercise 8. It is also useful to describe variability in the length of yellow perch. Why might this be important? Consider the following questions:

12. In several previous lessons, you described a data distribution using the five-number summary. Use the histogram and your answers to the questions in previous exercises to provide estimates of the values for the five-number summary for this sample:

    Minimum (min) value =

    Q1 value =

    Median =

    Q3 value =

    Maximum (max) value =

13. Based on the five-number summary, what is an estimate of the value of the interquartile range (IQR) for this data distribution?

14. Sketch a box plot representing the lengths of the yellow perch in this sample.

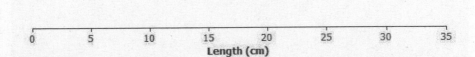

15. Which measure of center, the median or the mean, is closer to where the lengths of yellow perch tend to cluster?

16. What value would you report as a typical length for the yellow perch in this sample?

17. The mean absolute deviation (or MAD) or the interquartile range (IQR) is used to describe the variability in a data distribution. Which measure of variability would you use for this sample of perch? Explain your answer.

Lesson 20: Describing Center, Variability, and Shape of a Data Distribution from a Graphic Representation

©2015 Great Minds eureka-math.org
G6-M5M6-SE-B3-1.3.1-01.2016

Lesson Summary

Data distributions are usually described in terms of shape, center, and spread. Graphical displays such as histograms, dot plots, and box plots are used to assess the shape. Depending on the shape of a data distribution, different measures of center and variability are used to describe the distribution. For a distribution that is skewed, the median is used to describe a typical value, whereas the mean is used for distributions that are approximately symmetric. The IQR is used to describe variability for a skewed data distribution, while the MAD is used to describe variability for a distribution that is approximately symmetric.

Problem Set

Another sample of Great Lake yellow perch from a different lake was collected. A histogram of the lengths for the fish in this sample is shown below.

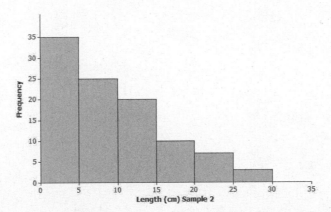

1. If the length of a yellow perch is an indicator of its age, how does this second sample differ from the sample you investigated in the exercises? Explain your answer.

2. Does this histogram represent a data distribution that is skewed or that is nearly symmetrical?

3. What measure of center would you use to describe a typical length of a yellow perch in this second sample? Explain your answer.

4. Assume the smallest perch caught was 2 centimeters in length, and the largest perch caught was 29 centimeters in length. Estimate the values in the five-number summary for this sample:

    Minimum (min) value =

    Q1 value =

    Median value =

    Q3 value =

    Maximum (max) value =

©2015 Great Minds eureka-math.org
G6-M5M6-SE-B3-1.3.1-01.2016

5. Based on the shape of this data distribution, do you think the mean length of a yellow perch from this second sample would be greater than, less than, or the same as your estimate of the median? Explain your answer.

6. Estimate the mean value of this data distribution.

7. What is your estimate of a typical length of a yellow perch in this sample? Did you use the mean length from Problem 5 for this estimate? Explain why or why not.

8. Would you use the MAD or the IQR to describe variability in the length of Great Lakes yellow perch in this sample? Estimate the value of the measure of variability that you selected.

Lesson 21: Summarizing a Data Distribution by Describing Center, Variability, and Shape

Classwork

Each of the lessons in this module is about data. What are data? What questions can be answered by data? How do you represent the data distribution so that you can understand and describe its shape? What does the shape tell us about how to summarize the data? What is a typical value of the data set? These and many other questions were part of your work in the exercises and investigations. There is still a lot to learn about what data tell us. You will continue to work with statistics and probability in Grades 7 and 8 and throughout high school, but you have already begun to see how to uncover the stories behind data.

When you started this module, the four steps used to carry out a statistical study were introduced.

Step 1: Pose a question that can be answered by data.

Step 2: Collect appropriate data.

Step 3: Summarize the data with graphs and numerical summaries.

Step 4: Answer the question posed in Step 1 using the numerical summaries and graphs.

In this lesson, you will carry out these steps using a given data set.

Exploratory Challenge: Annual Rainfall in the State of New York

The National Climate Data Center collects data throughout the United States that can be used to summarize the climate of a region. You can obtain climate data for a state, a city, a county, or a region. If you were interested in researching the climate in your area, what data would you collect? Explain why you think these data would be important in a statistical study of the climate in your area.

For this lesson, you will use yearly rainfall data for the state of New York that were compiled by the National Climate Data Center. The following data are the number of inches of rain (averaged over various locations in the state) for the years from 1983 to 2012 (30 years).

45	42	39	44	39	35	42	49	37	42	41	42	37	50	39
41	38	46	34	44	48	50	47	49	44	49	43	44	54	40

Use the four steps to carry out a statistical study using these data.

Step 1: Pose a question that can be answered by data.

What is a statistical question that you think can be answered with these data? Write your question in the template provided for this lesson.

Step 2: Collect appropriate data.

The data have already been collected for this lesson. How do you think these data were collected? Recall that the data are the number of inches of rain (averaged over various locations in the state) for the years from 1983 to 2012 (30 years). Write a summary of how you think the data were collected in the template for this lesson.

Step 3: Summarize the data with graphs and numerical summaries.

A good first step might be to summarize the data with a dot plot. What other graph might you construct? Construct a dot plot or another appropriate graph in the template for this lesson.

What numerical summaries will you calculate? What measure of center will you use to describe a typical value for these data? What measure of variability will you calculate and use to summarize the variability of the data? Calculate the numerical summaries, and write them in the template for this lesson.

Step 4: Answer your statistical question using the numerical summaries and graphs.

Write a summary that answers the question you posed in the template for this lesson.

©2015 Great Minds eureka-math.org
G6-M5M6-SE-B3-1.3.1-01.2016

Template for Lesson 21

Step 1: Pose a question that can be answered by data.

Step 2: Collect appropriate data.

Step 3: Summarize the data with graphs and numerical summaries.

Construct at least one graph of the data distribution. Calculate appropriate numerical summaries of the data. Also, indicate why you selected these summaries.

Step 4: Answer your statistical question using the numerical summaries and graphs.

©2015 Great Minds eureka-math.org
G6-M5M6-SE-B3-1.3.1-01.2016

Lesson Summary

Statistics is about using data to answer questions. The four steps used to carry out a statistical study include posing a question that can be answered by data, collecting appropriate data, summarizing the data with graphs and numerical summaries, and using the data, graphs, and summaries to answer the statistical question.

Problem Set

In Lesson 17, you posed a statistical question and created a plan to collect data to answer your question. You also constructed graphs and calculated numerical summaries of your data. Review the data collected and your summaries.

Based on directions from your teacher, create a poster or an outline for a presentation using your own data. On your poster, indicate your statistical question. Also, indicate a brief summary of how you collected your data based on the plan you proposed in Lesson 17. Include a graph that shows the shape of the data distribution, along with summary measures of center and variability. Finally, answer your statistical question based on the graphs and the numerical summaries.

Share the poster you will present in Lesson 22 with your teacher. If you are instructed to prepare an outline of the presentation, share your outline with your teacher.

Lesson 22: Presenting a Summary of a Statistical Project

Classwork

A statistical study involves the following four-step investigative process:

    Step 1:   Pose a question that can be answered by data.

    Step 2:   Collect appropriate data.

    Step 3:   Summarize the data with graphs and numerical summaries.

    Step 4:   Answer the question posed in Step 1 using the numerical summaries and graphs.

Now it is your turn to be a researcher and to present your own statistical study. In Lesson 17, you posed a statistical question, proposed a plan to collect data to answer the question, and collected the data. In Lesson 21, you created a poster or an outline of a presentation that included the following: the statistical question, the plan you used to collect the data, graphs and numerical summaries of the data, and an answer to the statistical question based on your data. Use the following table to organize your presentation.

Points to Consider:	Notes to Include in Your Presentation:
(1) Describe your statistical question.	
(2) Explain to your audience why you were interested in this question.	
(3) Explain the plan you used to collect the data.	
(4) Explain how you organized the data you collected.	

©2015 Great Minds eureka-math.org
G6-M5M6-SE-B3-1.3.1-01.2016

(5)	Explain the graphs you prepared for your presentation and why you made these graphs.	
(6)	Explain what measure of center and what measure of variability you selected to summarize your study. Explain why you selected these measures.	
(7)	Describe what you learned from the data. (Be sure to include an answer to the question from Step (1) on the previous page.)	

Lesson Summary

Statistics is about using data to answer questions. The four steps used to carry out a statistical study include posing a question that can be answered by data, collecting appropriate data, summarizing the data with graphs and numerical summaries, and using the data, graphs, and numerical summaries to answer the statistical question.

EUREKA MATH

Template for Lesson 22: Summarizing a Poster

Step 1:   What was the statistical question presented on this poster?

Step 2:   How were the data collected?

Step 3:   What graphs and numerical summaries were used to summarize data?

Describe at least one graph presented on the poster. (For example, was it a dot plot? What was represented on the scale?) What numerical summaries of the data were included (e.g., the mean or the median)? Also, indicate why these particular numerical summaries were selected.

Step 4:   Summarize the answer to the statistical question.

This page intentionally left blank